W9-BRR-905

A History of Tunnels

A HISTORY OF TUNNELS

Patrick Beaver

With diagrams by
AMÉ BEAVER

THE CITADEL PRESS
SECAUCUS, NEW JERSEY

First American edition, 1973
Copyright © 1972 by Patrick Beaver
All rights reserved
Published by Citadel Press, Inc.
A subsidiary of Lyle Stuart, Inc.
120 Enterprise Ave., Secaucus, N.J. 07094
Manufactured in the United States of America
Library of Congress catalog card number: 72-95415
ISBN 0-8065-0369-6

Contents

Illustrations

Diagrams

Acknowledgements

The author and publisher wish to thank the following for the willing help they have given:

D. G. Lloyd, F.I.C.E.
Bernard Howarth-Loomes
The South African Department of Water Affairs
The Orange–Fish Tunnel Consultants, Sir William Halcrow and Partners, London: Keeve Steyne and Partners, Johannesburg
Edmund Nuttall Limited
The Swiss National Tourist Office
London Transport
Holman Brothers Limited
M. T. Walters Associates Limited
K. M. Tunnelling Machines Limited
Ingersoll Rand

A History of Tunnels

Tunnelling: the Basic Operations

THERE ARE MANY reasons why animals dig into the earth: to find food and water; to make a home; to hide from danger; to store food. Man burrows for all these same reasons together with a few more of his own. It may be said with certainty that tunnelling was man's first exercise in engineering. The enlarging of the cave wherein he lived would be an obvious operation to a creature capable of logical thought. In fact, the remains of Stone Age victims of tunnel collapses have been found together with their tunnelling implements. The accidental discovery of water and mineral deposits during the operation of extending the home led men to bore for that purpose alone, and when tunnelling had developed from an inborn animal instinct to an art it was applied to as many purposes as it is today—which include, in addition to those listed above, drainage, sewerage, storage and the transport of men, materials and goods.

Tunnels may be 'shallow'—that is, a few feet below the ground, such as subways in cities; 'low-level'—as, for example, the London underground railway; or 'high-level', like those driven through hills and mountains. They may be of circular, oval or 'horseshoe' shape. Square or rectangular tunnels are not unknown, but these forms do not stand up well to the pressures exerted upon them by the adjacent ground. The engineer takes two main factors into consideration when determining the shape of a tunnel. First, its functional requirements; secondly, the type and condition of the ground. If, for instance, the vertical pressures are found to exceed the horizontal pressures, then the width should exceed the height and vice versa. But in practice a circular or near-circular shape is usually adopted as a compromise.

Rock and soil

The ground through which a tunnel is advanced can be broadly divided into two types—hard and soft. Hard ground may range from such rocks as granite and schist to limestone shale or sandstone. Soft ground can be subdivided into four categories.

Ground consisting of silt, sand, etc., either wet or dry, that has to be supported immediately and continuously.

Ground consisting of wet clay or earth or of such materials that will stay firm for a few minutes after excavation.

Ground comprising rocks and clays, that can be safely timbered well behind the excavation.

Ground that is self-supporting.

Initial operations

The most common method of driving a tunnel of any length is to first sink a number of shafts along the projected path. From the bottom of these, small pilot tunnels or headings are driven out in two directions. These are later opened out to size by one of a number of methods that will be described later. When the pilot passages are driven at the level of what will (later) become the roof, or 'arch', of the finished tunnel, it is called a 'top heading'; if it is at the level of the floor, or 'invert', it is known as a 'bottom heading'.

Sometimes a small 'pioneer' tunnel is driven parallel to and ahead of the main tunnel. At intervals a cross-heading is made to the path of the main tunnel to create two more working-faces. The advantages of the

←——— DIRECTION OF ADVANCE

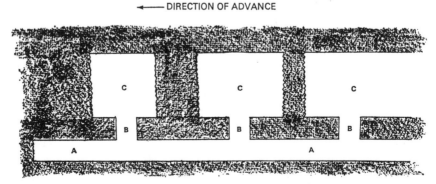

Figure 1. Pioneer method of tunnel driving.
A. 'Pioneer' tunnel. B. Cross-headings. C. Main tunnel.

'pioneer' system are many: it gives clear access for the removal of débris; it supplies a means of ventilating the main workings; it can be used as a drainage pipe; and, of course, it is always there should the main tunnel be later enlarged.

The organisation of the transport of men, tools, materials and débris to and from the working-faces of a long tunnel is of vital importance and has been likened to the circulation of blood in the human body. The least interruption will have far-reaching effects upon the rest of the workings. Again, the working space available is extremely confined and only one operation can be carried out at a time. Great attention is given, therefore, to timing a cycle of working operations. The following is a typical one, worked in driving the New Cascade Tunnel in the United States, in 1929.

7.10 A.M. A charge of 195 pounds of gelatin fired.

7.10–7.40 Powerful pumps force out air from the heading at 6,000 cubic feet per minute to clear away gas, dust and foul air.

7.40–8.40 Mucking-gang clear loose rock by means of small railway wagons.

8.40–9.20 Excavating gear is cleared away and drill carriage (with four drills) brought up to work-face.

9.20–10.35 Crew of eight drill seven holes in heading to a depth of 8 feet 6 inches, while the exact amount of explosive is brought up.

10.35–11.50 Half the drilling-crew take back the drilling carriage, while the others prime the holes with explosive and test and connect the firing leads. The heading is then evacuated by all except the single firer, who explodes the charges from the nearest cross-heading.

The whole cycle is then repeated.

In contrast to this, it has of recent years often been found necessary to deliberately retard the cycle of drilling, blasting and mucking because the advancing of these techniques can result in the tunnel face racing ahead of secondary work—lining, service installation and so on.

When once the line of a tunnel is decided upon, a thorough survey is undertaken to ascertain the nature of the ground to be traversed. Electronic devices inform the engineer of the presence of underground rivers, rock and clay; then, if possible, borings are made by means of a diamond drill. This consists of a hollow steel pipe with a diamond-faced cutting-edge that can penetrate the hardest of rock. As the drill rotates, a core of ground is forced through the pipe and examined by the geologist making the survey. When examining the ground for a tunnel under a mountain, say, these methods are not of course available, and the surveyor must deduce the nature of the ground from indications on the surface. These deductions, unfortunately, are not always an accurate indication of the conditions that will eventually be encountered.

If the ground which is to be tunnelled is accessible, the engineer can set his line on the surface with the aid of a surveying instrument known as a theodolite, which gives him bearings that are accurate to within 1/180th of a degree. He can then dig vertical shafts along the path of the tunnel to provide working-faces as well as ventilation. To check level and alignment between shafts the engineer may use plumb-lines. Two of these, about 10 feet apart, are hung down the shafts and aligned on the surface with a theodolite. At the bottom of the shaft a bearing is taken along these lines that gives an accurate course for the tunnel. When the path of a proposed tunnel passes under inaccessible ground, such as a mountain, where for obvious reasons shafts cannot be sunk, the theodolite is used inside the tunnel where it is 'sighted back' to its first position in the portal. It is then turned through 180° so that it points exactly forward, indicating quite accurately the line to be tunnelled. This method of lining up is so exact that engineers are surprised if there is an error of more than a fraction of an inch when the two headings meet.

In the case of the 5,000-foot-long Musconetcong Tunnel in New Jersey, for instance, the two paths were only half an inch out of alignment, and one-sixth of an inch off level. In the 25,000-foot Hoosac Tunnel in Massachusetts, that took twenty-five years to complete, the errors were even less; while in the famous Alpine tunnel under Mont Cenis, which was a severe trial to all concerned, the error in direction was nil.

Excavation systems

Except when driving through hard rock, the tunnelling engineer is faced with two basic problems: to support the roof, face and sides of his heading between the operations of excavating and lining it; and to carry out the various operations of tunnelling—for example, excavating, timbering, mucking and lining—in the necessarily confined working space that the tunnel offers. As a general rule, it may be said that driving through very soft ground is the most difficult form of tunnelling. The excavation must be supported in some manner as soon as it is completed—otherwise the walls may cave in or squeeze together, the roof will tend to fall and even the bottom may bulge upwards. These sliding movements may exert enormous pressures upon timbering and can even crush it completely. Nowadays, all major tunnelling operations through soft ground are carried out using the shield method which does not require timbering. But it was not always so; a variety of methods were evolved during the last century, most of them taking their name from the country in which they originated.

The Belgium method

This method was first employed in building the Charleroy tunnel on the Brussels–Charleroy canal, in 1828.

The excavation is begun by driving a top-centre heading (*see* Fig. 2, section 1); and this is carried ahead for a certain distance, which depends upon the nature of the soil. This top heading is immediately timbered.

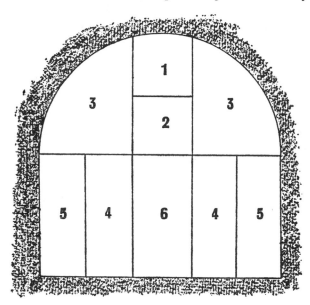

Figure 2. Belgium method—order of excavation.

Working behind the leading work-party, another gang deepens the top heading to a depth corresponding to the springing lines of the arch (section 2). Farther back still, the two upper side sections, or 'haunches', are removed by four gangs working at the two sides and the two fronts (sections 3). When the entire upper part (sections 1, 2 and 3) has been completely excavated, the roof arch is built, its feet supported by the un-excavated sections (sections 5). Sections 4 are now removed to leave two transverse trenches and, in these, struts are inserted to underpin the masonry arch (*see* Fig. 3). When sections 5 have been excavated, the arch can be completed from the feet up and then the timbering can be removed. Finally the centre bench (section 6) is removed.

The efficiency of the Belgium method of construction depends greatly upon choosing such dimensions for the various sections that each can be excavated at about the same speed.

The advantages of the method are that (1) the excavation advances at several points at the same time without the various gangs getting in each other's way; (2) the excavation consists of a number of drifts, or sections, which are immediately strutted, thus allowing no time for the surrounding ground to shift; and (3) the roof of the tunnel, which is subject to the greatest pressures, is built first.

Figure 3. Belgium method—timbering and underpinning of the arch.

The disadvantages of the method are that (1) the roof arch is liable to sink, especially in very soft ground; and (2) before the floor, or invert, is built there is a danger of the walls sliding under pressure.

The German method

This system (*see* Figs. 4–6) was first used in 1803 to build the St Quentin canal tunnel. It is also known as the 'centre-core method', since its underlying principle is to leave a central bench of ground to be excavated last and to use it to support roof and wall timbering, in order that the arching can be built in one operation—unlike the Belgium method which has the disadvantage of building arch and walls separately.

First, two side headings are dug (*see* Fig. 4, sections 1), each about 7 feet wide and one-third the total height of the finished gallery; these headings

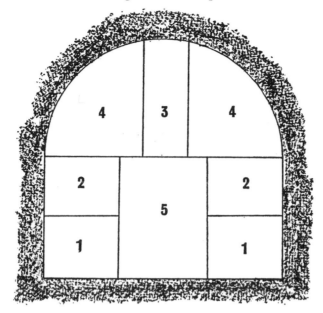

Figure 4. German method—order of excavation.

Figure 5. German method—timbering supported by centre core.

must be large enough for the construction of timbering and masonry as well as for the passage of skips to remove the excavated spoil. Two work-gangs following behind increase the height of the two side headings to the springing points of the arch (sections 2). Farther back again, a top-centre

Figure 6. German method—arch during construction, showing falsework for arch masonary.

heading is driven (section 3) which, in turn, is broken out to form the two haunch headings (sections 4) and joins up with the side headings (sections 1 and 2). The still unexcavated centre-core (section 5) is then used as a support for the wall, roof timbers and falsework until the arch is completed; it is then dug out.

Modified German method

If, during tunnelling through fairly firm ground, water was encountered, the sequence of excavation of the German system was modified as follows.

A small bottom drift was first excavated to act as a drain (*see* Fig. 7, section 1). The top-centre heading was then driven (section 2). The haunch headings (sections 3 and 4) were then dug out in four sections, and the side sections (5 and 6) sunk to grade. This left the centre-core (section 7) to bear the timbering until the lining was built.

The German system claimed many advantages, the chief of which was the use of the core to save timbering, but its defects were many. Hauling the skips through such small spaces tended seriously to interfere with the carpenters and masons working behind the face, thus increasing labour

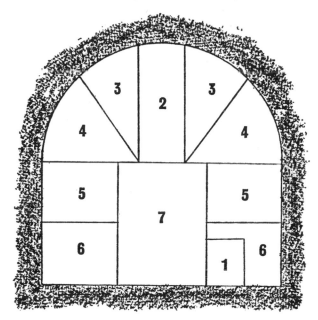

Figure 7. Modified German method—order of excavation.

costs. Ventilation was bad; faulty bonding was often the result of building masonry in cramped spaces and, finally, it was found to be very difficult to preserve alignment in a tunnel so constructed.

The English method

This method received its name from being used to tunnel through the kind of ground usually encountered in England—clay, shale and sandstone. Its main characteristic was the excavation of the full tunnel in one operation—a length being dug out and lined before the next advance was made. Using this method it was usual first to drive a small-bore pilot heading to establish the axis of the tunnel. Whether this pilot was situated at the top or the bottom was entirely at the discretion of the engineer; its position made no difference to the subsequent mode of excavating the tunnel proper. Work on the main heading always began at the upper part (*see* Fig. 8, section 1).

When the top-centre heading had advanced a certain distance, usually between 12 and 20 feet, it was enlarged, step by step, sideways and strutted with temporary posts resting on blocks and supporting longitudinal bars. The widening was continued until the full section (sections 2) was excavated. Along the sides of this top heading, a heavy sill was laid transversely and permanent struts erected from it to the longitudinal bars,

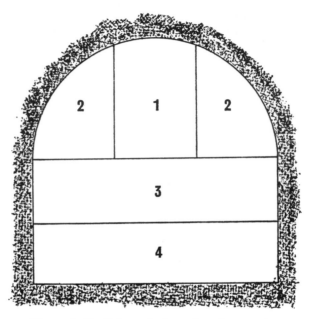

Figure 8. English method—order of excavation.

the temporary work then being cleared away. Section 3 was dug out by making a centre-trench through it that was then widened either side, temporary posts being inserted to support the sill in sections 1 and 2. A second transverse sill was then placed in section 3, below the sill in sections 1 and 2, and struts placed between them. The excavation of section 4 then proceeded in exactly the same manner as section 3, and the completed length of tunnel was lined with masonry.

The great advantage of the English method was that the masonry was built in one piece from foundation to arch and resulted in a strong, homogeneous construction. Its disadvantage was that tunnellers and masons worked in turn, thus making the method the most expensive of all. The English method was, in Europe, almost entirely confined to its country of origin, but it found favour in the United States, being used to build the Hoosac, Musconetcong, Allegheny, Baltimore and Potomac, as well as many other tunnels in America.

The Austrian method

This method was first used in constructing the Oberau tunnel on the Leipzig and Dresden Railway, in 1837, and it resembles the English in so far as the full section is excavated and the masonry is built up from the foundations. The essential difference between the two systems is that the

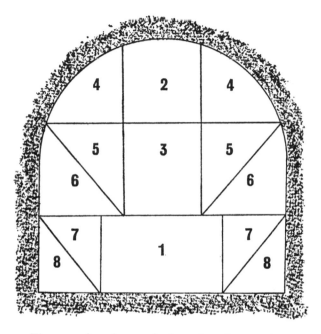

Figure 9. Austrian method—order of excavation.

excavation is carried out in small sections and is continued; in the English method excavation alternates with masonry construction. A centre-bottom heading (*see* Fig. 9, section 1) was first driven, rising to at least half the height of the springing lines of the arch. When this had advanced for a distance of about 15 feet the centre-top heading (section 2) was driven for the same distance. Section 3 was then removed by men working from the top heading. This facilitated the removal of spoil which was dropped straight down into skips in the bottom heading (section 1). The result was a central passage, the full height of the tunnel. This passage was opened out to full section by removing the remaining parts in numerical order, the strutting being inserted as shown in Figures 10–10b.

The masonry lining was built up from the foundations of the side walls to the crown of the arch in consecutive sections of about 15 feet in length.

The advantages of this method were that (1) the driving of a large number of small headings which were immediately strutted allowed for no disturbance of the surrounding ground; (2) the masonry was built from the

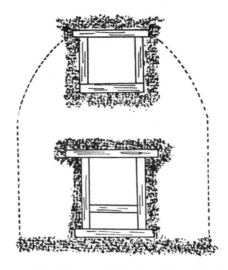

Figure 10. Austrian method—insertion of timbering.

foundations upward; and (3) masons and excavators did not interfere with one another. The main disadvantage was that the strutting was liable to distort or give way under unsymmetrical pressures.

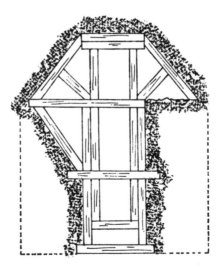

Figure 10a.

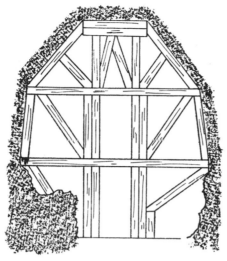

Figure 10b.

The Italian method

This method was first used in the building of the Cristina tunnel in Italy (*see* Fig. 11). Particularly suited to penetrating soft and treacherous soil, the system is unique in that the bottom half of the section is excavated and then filled in again later. It is essentially a method for excavating narrow tunnels. Applied to railway tunnels for instance it was only used for single tracks, since any ground that was treacherous enough to justify such an expensive method of tunnelling was not suitable for a heading of sufficient size to take a double track unless a shield was used.

A centre-bottom heading was first driven (section 1), which was then broken out to full width (sections 2). The invert and side walls were then built and the excavation filled in again with earth. A centre-top heading (section 3) was then driven and opened out by removing the soil in section 4. The walls of section 4 were inclined to reduce any tendency to slide and to allow a maximum number of struts to be placed. The haunches (sections 5) were then excavated, and when this was done the entire tunnel had been dug with the exception of the strip (section 6) which rested upon the refilled sections (1 and 2). At the ends of section 6, trenches were dug to reach the tops of the side walls already built in the lower (filled in) half of the tunnel. The masonry was built up and completed, the remainder of section 6 excavated and the earth filling in the lower sections was removed. Headings were usually driven from 6 to 10 feet ahead of the lining.

We have seen that the principle underlying all these methods was the

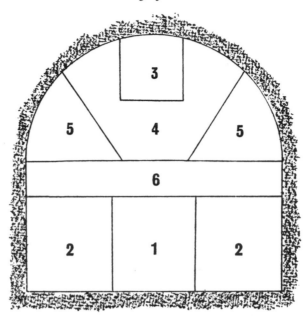

Figure 11. Italian method—order of excavation.

driving of one or more small headings that were later enlarged and joined together. When these small headings were driven through soft soil a system known as 'fore-poling' was used. There were many variations of fore-poling, but it will be sufficient to describe one of these.

If we assume that the 'poling-boards' (*see* Fig. 12, a and b) are in

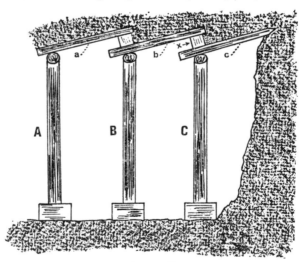

Figure 12. Fore-poling system.

position and are supported by square frames (A, B, C), the next step would be to insert the poling-board 'c' over the crown bar of the frame C, and *under* the block 'x'. Soil would then be removed from the top of the work-face and as the excavation advanced, the poling-board 'c', which had a sharpened end, was driven ahead until its rear end only slightly overhung the crown bar of frame C. The remainder of the face of the heading was then removed until it nearly reached up to the *front* end of the poling-board 'c'. Another frame,was then set up and the process repeated. Poling-boards were placed in the sides of the heading in a similar manner.

Whether tunnelling through soft ground or hard, by whatever method, it is usual to drive a heading in advance of the main excavation for the following purposes: (1) to fix, correctly, the axis of the tunnel; (2) to be forewarned of the nature of the material to be encountered, and to be ready to deal with any troubles caused by a change in the soil; and (3) to draw off any water that might be encountered.

In very long tunnels through rock, driven from two ends, the pilot headings are kept as much as 2,000 feet in advance in order that communication between the two workings may be made as soon as possible and the alignment verified; also, that as great an area of rock as possible may be dealt with in the work of breaking out. In short tunnels, where there is less liability to errors in level and alignment, shorter advance headings are used, while in soft ground the rule is the softer the ground the shorter the advance.

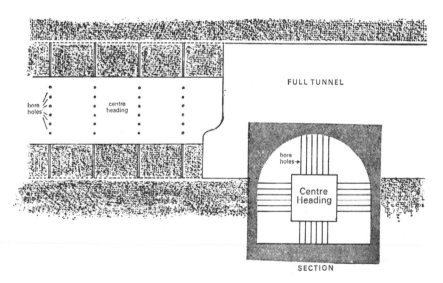

Figure 13. Radial method—longitudinal section.

Radial method

In tunnelling through hard rock the 'Radial method' may be used (*see* Fig. 13). This consists of a centre heading from which blasting holes are drilled outward. Explosives are then used to enlarge the tunnel to full size.

The upraise method

Another method of tunnelling through rock (*see* Figs. 14–14b) is first to excavate a lower heading for several yards; along this is laid a narrow-gauge railway. Vertical shafts, or upraises, are then driven through the roof of the bottom heading and from these a top heading is driven in two directions leaving a dividing floor of a few feet between the two headings. The upraises are used as chutes to load spoil directly from the top heading to the skips below. As the headings progress the top one is enlarged and widened to form an arch. The dividing floor is then removed and the lower heading broken out to full size.

Hazards and casualties

Tunnelling has always been a hazardous occupation and is likely to remain so for a long time to come. Even today, with all the scientific aids available to him, the tunnelling engineer cannot be absolutely certain of what awaits him underground; indeed it is not infrequent for the direction of a tunnel to be changed while it is being built. The overriding factor in tunnelling is, of course, the nature of the ground along the tunnel's projected path. This can vary from almost liquid mud to hard, unfissured rock. Fissures in hard rock can render it as hazardous to penetrate as soft ground, for they may contain water under extremely high pressure which, when broken into, can burst into a heading at the rate of thousands of gallons a minute. Under mountains a tunnel may be subjected to squeezing forces that are nigh on irresistible, crushing to splinters even the most massive timbers. Dry clay is another tunnelling hazard, for when exposed to the atmosphere it absorbs water and expands, creating enormous pressures. An example of this occurred in the Tanna tunnel in Japan, where it was necessary to line the tunnel with 6 feet of masonry to hold back the clay. Pockets of methane gas constitute a further danger which was particularly severe in the days when the work was lit by the light of candles.

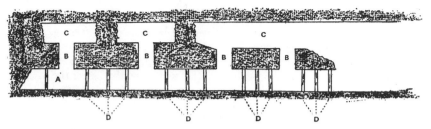

Figure 14. Upraise method—longitudinal section.
A. Bottom heading. B. Upraises. C. Top heading. D. Temporary timbering to support dividing (centre) section.

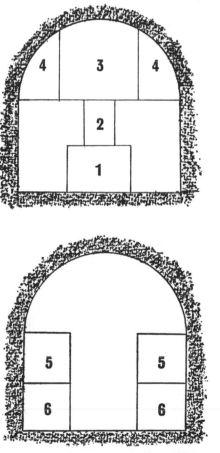

Figures 14a, b. Upraise method—order of excavation.

The trials and cares of the tunnelling engineer are summed up as follows by a nineteenth-century writer:

> In tunnelling . . . more so than in any other branch of engineering, it is the engineer's function to direct the work, not only in general terms, but even down to the smallest details. Whoever has been privileged to have charge of the construction of a shield-driven tunnel will know that, after the location has been fixed, the design made and work started, he will be charged with a never-ceasing load of cares that cannot be left to the contractor, however proficient, efficient or experienced the latter may be. To start with, every move of the shield is in itself an experiment which must be controlled by the engineer. Every change of ground demands a change in the methods of construction. It is the engineer's duty to know beforehand where these changes will occur and to be prepared for them and not merely wait for them to show themselves. It is he who must direct what exploratory borings must be made to predetermine these changes of ground, and it is he who must know what methods are best suited to the conditions thus disclosed. He must have knowledge of the proper air pressure to carry under a given set of conditions; he must know where the bulkheads should be built; he must be prepared, in fact, to give to every phase of the work the most broad and basic directions and at the same time to see that in the minutest details the work is conducted so as to give the best possible results that the circumstances will allow. Heavy indeed is his load of care. He owes it to his clients and equally to his own conscience that his location, design and direction are such that the work is sound and permanent, built with due economy of money, time and labour and that the lives of those who are carrying out the work may not be endangered by collapse, flood or disease.

But even the greatest care on the part of the engineer cannot prevent accidents during work that is fraught with so many risks. Over-familiarity with these dangers can breed a carelessness among workmen that is, more often than not, the cause of many of the deaths and injuries that occur. The experience of engineer H. E. C. Cowie in India, in 1897, is a good example of this. Six of his men, all experienced miners, were killed during the course of one year while constructing a short tunnel on the Mari-Attock Railway in the Punjab. The tunnel was through soft sandstone—one of the easier materials through which to tunnel—and Cowie was using black powder for blasting. He gives the following list of fatal casualties that occurred in the 1,000-foot tunnel.

1. Two men were killed through carelessness when lining the bottom heading.
2. One man was killed by a lump of rock falling from the roof. This occurred through neglect of the order to trim the excavation with pickaxes, an operation that detects loose pieces.

3. A powder explosion occurred from a Pathan miner lighting his pipe over the powder canister! Both pipe and canister were strictly forbidden inside the tunnel. A stampede among the masons resulted, who, in the dark, jumped down from the centring. Two were killed and many injured. The guilty Pathan, says Cowie, 'came out scorched and smiling'.

4. One man was killed by an insecurely placed prop which fell on him after a blast.

So numerous and deadly are the hazards of tunnelling that perhaps it is as well that most of the major tunnelling operations have been journeys into the unknown; had they been known by the engineers, many of them—the Simplon for example—would never have been started.

Fire, suffocation, drowning, falling rock and pockets of gas have taken their underground toll of human life for centuries, and when explosives were added to the tunnellers' equipment other hazards also came. A terrible accident occurred in 1909, during the excavation of one of Chicago's water tunnels. One hundred men were at work on a wooden crib in deep water. The crib caught fire and sixty of the men died through burning, drowning or suffocating.

Pockets of sulphurated hydrogen constitute another danger. This gas, when inhaled, can cause severe illness as well as blindness, while the highly explosive gas, methane, can be ignited by a spark from pick or shovel.

The greatest and most common danger is, as we shall see, water, which can suddenly appear in enormous quantities no matter what type of ground is being traversed. Describing the building of the 3½-mile Totley tunnel in the 1890s, the *Manchester Guardian* remarked that 'every man seemed to possess the miraculous power of Moses, for whenever a rock was struck, water sprang out of it'.

The greatest underground killer of all was, of course, the diseases to which tunnellers were particularly prone. The number of deaths from silicosis and tuberculosis must, over the centuries, have been vast and, of course, they would not be recorded as the toll of tunnelling because the victims would have lingered long after the ailment rendered them too weak for the heavy work demanded underground.

The conditions under which the tunneller may be called upon to work are severe indeed and any number of dangers and discomforts may be present at the same time. He knows full well that to bore through a mountain involves the disturbance of natural balances that were formed millions of years ago, and the tunneller cannot complain if the effect of these disturbances are the cause of the most perilous phenomena. The squeezing forces of the mountain above him may cause jagged slabs of rock, a ton in weight or more, to 'spall' or fly off at high speeds and dangerous angles. Pockets of underground gas may half blind him or may set fire at the smallest spark; water or mud may suddenly burst in upon him, filling the tunnel before he can escape. Always there is the danger of a roof fall.

Then there is the frightful roar of twelve or more rock drills simultaneously biting into hard stone in the echoing confines of a heading, the dust and smoke of the atmosphere and the lack of *certain* knowledge of what lies ahead. Even today with all the aids that science has brought to the art of the tunneller, the average cost in life is one to every mile of tunnel. And yet to most men who work in tunnels, no rock is too hard, no soil too soft, no ground too dangerous and no conditions too severe for them to tackle. They adjust themselves to the situation in hand and shove, push, batter and slog their way through whatever they encounter. Their only thought is to push ahead and, if possible, create new tunnel-driving records.

2

Tunnels of the Ancient World

THE DOGGED PERSEVERANCE shown by the ancients in their tunnelling operations is astonishing, and especially so when we consider the primitive tools and methods they employed. Tunnelling through soft ground with hand tools, the simplest timbering and without adequate ventilation must have been formidable a task enough, while the driving of long tunnels through iron-hard rock with little else but hammer, chisel and brute strength is a telling indication of what man can do, and does, when his mind is set on it.

Although the brute-strength-and-sweat method of tunnelling was employed up to comparatively recent times, it was the Ancient Egyptians who first applied science with a view to lessening its labour. These methods were later copied by the engineers of ancient Greece and Rome and were used throughout Europe until the introduction of gunpowder for blasting rock.

Fire-setting

To loosen rock for removal, the ancients cut holes around large blocks and achieved the final breaking by driving wedges or by inserting wooden plugs between the blocks which, on being soaked with water, swelled up to break the rock away. Fire-setting was another ancient method of rock tunnelling. This consisted of heating the rock and suddenly cooling it with the result that it disintegrated. Pliny says, 'Vinegar poured upon rocks in considerable quantities has the effect of splitting them, when the action of fire alone has been unable to produce any effect thereon.' The Roman engineer Vitruvius agreed, but said that 'considerable quantities' of water would have produced the same effect.

Conditions underground during a fire-setting operation must have resembled those described in Dante's '*Inferno*'. In addition to the choking smoke of the confined fire and the hot steam from water or vinegar, the cooling operation would have generated poisonous vapours from some forms of rock. The fumes from sulphureous or arsenious ores, for example, must have caused the most fearful suffering and the deaths of countless labourers. But then, the essential ingredient of much ancient engineering was an unlimited source of manpower in dispensable form, and the nations that could afford great construction projects always possessed a good supply of slaves and captives.

Fire-setting as a means of tunnelling through hard rock was used, although rarely, up to the end of the nineteenth century, the method being to employ a large furnace on wheels, so arranged that a strong draught passed out at the front and threw the flames directly against the rock face. During that same period the old fire system was used to enlarge headings and chambers in exceptionally hard rock—especially in situations where it was desired to bring down the roof. Some engineers argued that it was easier and cheaper in large tunnels, driven with a bottom heading requiring no timbering, to use the fire system to break out upwards. But even if this system was a practical one there remained the danger of shattering the roof which, by ordinary driving, would have been perfectly safe.

The Egyptians

As in other forms of civil engineering the Egyptians led the world in the art of tunnelling by many centuries. On ascending the throne an Egyptian king set slaves to work excavating a tunnel into rock, the end of which would be reached on his death and which then became his tomb. The tunnel tomb of Mineptah, at Thebes, was driven into a hill of rock for a distance of 350 feet. There a shaft was sunk and from its bottom the tunnel continues for a further 300 feet before opening out into the king's tomb. The Egyptian tunnellers used copper saws and reed drills supplied with emery dust and water; they employed tube drills that left a core of rock exactly like that of a modern diamond drill. The Egyptians also tunnelled long conduits and underground chambers for storing water during the dry season.

At Abu-Simbel on the Nile in Upper Egypt are the temples of Rameses II, hewn out of solid rock to a distance of 150 feet. The largest of them all, the tomb of Seti I, has a total length of some 700 feet. One of the oldest inscriptions in the world, dating from the tenth century B.C., refers to a tunnel driven for the purpose of supplying water to a town. After describing a successful rebellion against Israel by Mesha, King of Moab, the inscription reads: 'Mesha then built two conduits and as the city of Karcha possessed no cisterns he ordered the inhabitants of the city to place a cistern in every house. He then set the Israelite prisoners to constructing a conduit to bring water to the cisterns.'

In the year 715 B.C., King Hezekiah, fearing that the Assyrians might besiege Jerusalem, built a tunnel to carry water from the Pool of Virgins to the Pool of Siloam which lay within the city's fortifications. The two pools are only 110 feet apart but the tunnel was much longer, running in a curve like the letter 'S'. An inscription at the Siloam end said the tunnel was driven from both ends and that the workmen, hearing each other's voices, were able to change direction in order to meet. This would explain the curious path the tunnel took.

> Behold the excavation. This is the chronicle of the excavation. While the workmen were still digging with their axes towards one another with three cubits still separating them, each heard the voice of the other and they called to each other. Still was there an excess of rock on the right and on the left. On the day of the excavation the workmen struck out each in the direction of his neighbour and they met, axe upon axe. And the waters flowed from the spring to the pool a distance of a thousand and two hundred cubits.

Mention of Hezekiah's tunnel can be found in II *Kings* xx 20, and the fact that it is listed among other works of engineering suggests that tunnelling was not considered remarkable in biblical times.

A water tunnel built in the year 687 B.C. on the Greek island of Samos, was described by Herodotus as 'one of the greatest works to be seen in the Greek world'. This tunnel is the underground section of a water conduit that for the most part is in the open. Herodotus describes the tunnel as a 'double-mouthed channel pierced . . . through the base of a high hill; the whole channel is seven furlongs [seven-eighths of a mile], eight feet high and eight feet wide; and throughout the whole of its length there runs another channel twenty cubits deep and three feet wide through which the water coming from an abundant spring is carried by its pipes to the city of Samos. The designer of this work was Eupalinos, son of Naustrophos, a Megarian.' The tunnel was dug from both ends to meet in the middle, but the two teams missed each other by some 16 feet and the meeting was made by way of a right-angle bend. The entire work, which took many years to complete, was done with hammers and chisels.

The Indians

The world's most beautiful and elaborate tunnels are to be found in the rock temples of India. Built to imitate the earlier wooden temples, the rock walls of the tunnels are ornamented with all the florid embellishments of Indian wood-carving. It should be noted that in all this most beautiful tunnel work there is not a particle of masonry or stucco; all was cut out of the hardest rock with simple hand tools. The tunnels of Ellora alone add up to 6 miles length, all cut with hammer and chisel.

Tunnels in warfare and siege

The use of tunnels in warfare dates from the very earliest times as a most effective method of tumbling the walls of besieged cities. A tunnel could be started at a place well out of the sight and range of the weapons of the besieged or from behind an earthwork. Lined throughout with timber, it could be brought up to, and under, the walls of the city where it would then be enlarged. When the tunnellers had withdrawn, the timber lining would be set alight bringing down the roof of the tunnel together with a section of the city walls. There can be little doubt that the walls of Jericho were destroyed in this way. The fanfare of trumpets probably being the signal for the besieging army to advance at the moment that the walls collapsed, or part of a diversionary operation to draw the defenders away from the undermined section.

However, the besiegers did not always have things their own way. Describing the siege of Appolonia in 214 B.C., the Roman architect Vitruvius writes[1]:

> Again, when Appolonia was besieged, and the enemy designed by digging tunnels to penetrate unsuspected within the walls, this was reported by spies to the citizens of Appolonia. They were panic-stricken at the news and their spirits failed them in their lack of resource. For they did not know the time or the place for certain where the enemy were likely to emerge.

> But at that time Trypho of Alexandria was the architect in charge. Within the walls he planned tunnels and, removing the soil, advanced beyond the wall to a distance of a bowshot. Everywhere [along the tunnel] he hung bronze vessels. Hence in one excavation which was over against the tunnel of the enemy, the hanging vases began to vibrate in response to the blows of iron tools. Hereby it was perceived in what quarter their adversaries

[1] Vitruvius, *On Architecture*, Book X, translated by Frank Granger (Heinemann, London 1931).

purposed to make an entrance with their tunnel. On learning the direction, he filled the bronze buckets with boiling water and pitch overhead where the enemy were, along with human dung and sand roasted to a fiery heat. Then in the night he pierced many openings, and suddenly flooding them, killed all the enemy who were at work there.

The addition of the dung seems to have had no other purpose than to add a nasty insult to what must have been very nasty injuries. Vitruvius also describes the routing of attacking tunnellers at the siege of Marseilles in 49 B.C.:

Again, when Marseilles was besieged and the enemy drove more than thirty tunnels, the inhabitants were on their guard, and made a deeper ditch than the one in front of the ramparts. Hence all the tunnels came out into this. But where inside the wall a ditch could not be made, they dug a moat, like a fish pond, of great length and depth, over against the quarter where the tunnels were being made, and filled it from the wells and from the harbour. Hence when a tunnel had its passage suddenly opened, a strong rush of water flowed in and threw down the props. The troops within were overwhelmed by the collapse of the tunnel and the flood of water.

In 406 B.C. the Romans besieged the Etruscan town of Veii, the last of the Etruscan strongholds to resist the Roman conquest. The town held out for no less than ten years by means of a secret tunnel through which it received quantities of food, water and military supplies. In 396 B.C. the commander of the Roman army tunnelled under the walls of the town and into its centre; then, while an attack was mounted on the walls from the outside, a troop of Romans entered the city by way of the tunnel and attacked the Etruscans in the rear.

This was more than bad luck for the Etruscans, since it was they who had taught the art of tunnelling to the Romans. The latter, however, developed it and became the greatest tunnellers of ancient times. They built tunnels for thoroughfares, tunnels for water supplies, tunnels for drainage. They burrowed through rock and through earth wherever their conquests took them; and, along with the remains of their other great works, their tunnels can be seen today with their cement and masonry in excellent condition.

The aqueducts of the Romans

In 359 B.C. a tunnel was constructed to tap off the water of Alban Lake, which lies in the crater of an extinct volcano some 16 miles south of Rome. Livy says it was driven at the instance of the oracle of Delphi that prophesied that the Romans would never take Veii until they drained Alban Lake. It seems doubtful that the Romans would undertake such a great task of civil engineering on the strength of such a prophecy. A tunnel *did* have to be constructed to shift the defenders of Veii as we have seen, and writing some 400 years after the event the historian may have confused his tunnels. Cicero says that 'the work was intended to benefit suburban farms', and so it is possible that the lowering of the water was done to prevent it spilling over the side of the crater and flooding the surrounding farmland. Driven through the hardest volcanic rock the tunnel was 6,000 feet long, 5 feet wide and, on average, 8 feet high. The work took a year and was executed entirely by hand with 1-inch chisels. We are told that the tunnel was driven in two directions from a central shaft, but this is unlikely. Even if we allow that each of the two work-faces were penetrated at the impossible rate of $3\frac{1}{2}$ inches per hour—a galloping rate for hand chiselling through hard lava—then the work would have taken fourteen months if round-the-clock shifts were worked. But as it is known that some fifty shafts were sunk along the line of the tunnel, it would appear much more likely that the work was done on some hundred separate headings. Built at the level at which it was intended to lower the water—that is, 430 feet below the lip of the crater—it was not possible to cut a portal through into the lake. Instead, a vertical shaft was sunk near the edge of the lake and down to the level of the tunnel. When the tunnel was complete a heading was driven from the shaft to just below the level of the water which then went down. The process was repeated again and again until the level of the lake was the same as that of the tunnel.

Another outstanding Roman (road) tunnel was the one that carried traffic through the Posilipo Hill between Naples and Ponzzuoli. Driven through a calcium volcanic deposit known as tufa, the tunnel was 3,000 feet long and 25 feet wide. To admit light to the interior, the roof and floor converged towards the middle from the portals, which are no less than 75 feet high; but this device seems not to have worked, for Seneca, after traversing the Posilipo tunnel, commented:

> There is no prison longer than [the Posilipo tunnel] . . . no torch dimmer than those they shielded before us, which served not to lighten the darkness but only allowed us to look at one another. And in any case even had there been a little light it would have been hidden by the dust which was thick enough to darken an outdoors place notwithstanding the wind. What then can be said of dust in a place where it turns back upon itself and unstirred by wind, falls back upon those who raise it?

The date of the construction of the Posilipo tunnel is not known for certain but it is thought to have been built during the time of Augustus Caesar (63 B.C.–A.D. 14).

The building of another great water tunnel was undertaken by Roman engineers in A.D. 41 to drain Lake Fucino into the River Liris. Described by Pliny as the greatest public work of all time it was one of the wonders of the ancient world. Three and a half miles long and 20 feet high it was driven from some forty shafts sunk on the line—some to a depth of 400 feet—and occupied the labours of 30,000 men for over eleven years! Half a million cubic feet of hard rock was chipped from the Fucinus tunnel and hoisted up the shafts in copper buckets. The completion of this mighty undertaking made 50,000 acres of land available for farming.

The underground aqueduct that bears the name of its builder, the Emperor Hadrian, was driven to supply ancient Athens with fresh water from the underground springs of Mount Pentelikon and Mount Parnes. Built by the Romans during their occupation of Greece, it served the city for many years until, during the period of the Turkish occupation, the population of Athens fell to about 5,000 persons. Then the aqueduct became disused. For some 1,600 years it lay forgotten until, in 1840, it was rediscovered, repaired and put back into use. In 1925 the American firm of engineers, Ulen & Co., were employed to reconstruct the Hadrian aqueduct and to connect it with a new tunnel which it was to build. And so today the Roman tunnel still supplies water to Athens.

As originally built, Hadrian's aqueduct was more than 15 miles long and was driven from shafts at intervals of about 120 feet. The headings, $2\frac{1}{2}$ feet wide by $5\frac{1}{2}$ feet high, were just large enough to allow one man to work in each. Alignment was determined by dropping pairs of plumb-lines down neighbouring shafts and visually sighting along them. As the shafts were up to 130 feet deep the slightest deviation or error in a plumb-line of that length caused considerable errors in alignment. Consequently the headings deviated from true course very frequently and on approaching each other the two working parties corrected their courses with sound signals.

The accuracy attained by the classical engineers in their building works is amazing when we consider the primitive instruments at their disposal. Among those tools much relied upon by the Romans for levelling were the *libra aquaria* and *dioptra*, although it is not clear today what these instruments were. Another levelling device, the *chorabates*, consisted of a wooden plank some 20 feet long, mounted at either end on two legs of equal length. The legs were connected by diagonal braces on which were marked vertical lines. A plumb-line corresponded with these lines when the instrument was level. When the wind prevented the plumb-bobs from remaining stationary, channels in the upper surface of the plank were filled with water, and when the water touched the ends the level was taken to be correct. And so in this way observation of the elevation or descent of the ground could be accurately made.

In the year A.D. 148 the Roman engineer, Nonius Datus, was instructed to build an underground aqueduct at the town of Saldae in Algeria, and, it seems, he made the fundamental engineering error of leaving the contractors to get on with the job, without qualified supervision, while he travelled back to Italy for a stay of four years. In the year 152 the Roman governor of Saldae, Varius Clemens, summoned Rome 'to dispatch at once the hydraulic engineer of the third legion, Nonius Datus, with orders that he finish the work which he seems to have forgotten'. The engineer was duly dispatched to Algeria where, after inspecting the progress of the work on the tunnel, he reported to the magistrates of Saldae as follows.

> After leaving my quarters I met with brigands on my way, who robbed me of even my clothes, and wounded me severely. I succeeded, after the encounter, in reaching Saldae, where I was met by the governor, who, after allowing me some rest, took me to the tunnel. There I found everybody sad and despondent; they had given up all hopes that the two opposite ends of the tunnel would meet, because each section had already been excavated beyond the middle of the mountain, and the junction had not yet been effected. As always happens in these cases the fault was attributed to the engineer, as though he had not taken all precautions to ensure the success of the work. What could I have done better?
>
> I began by demarking the outlying spurs of the mountain and traced on the mountain ridge the axis of the tunnel as precisely as possible; I drew plans and cross sections of the whole work and handed them over to Petronius Celer, the governor of Mauritania; and to take extra precaution I summoned the contractor and his workmen and myself began the excavation in their presence, with the help of two gangs of experienced tunnellers; namely a detachment of infantry and a detachment of Alpine troops. What more could I have done? Well, during the four years I was absent at Lambaese, expecting every day to hear the good tidings of the water at Saldae, the contractor and his men committed mistake after mistake; in both the galleries of the tunnel they diverged from the straight, both to the right, and had I arrived a little later, Saldae would have possessed two tunnels instead of one.

After making a new survey the engineer caused the two parallel tunnels to be joined by a transverse tunnel. Eventually the waters of the river passed beneath the mountain to arrive at Saldae amid extraordinary, if belated, rejoicings.

Frontinus

The most prolific of the Roman tunnel builders was Sextus Julius Frontinus who was appointed Curator Aquarum, or chief water engineer, of Rome in A.D. 95 when he was sixty years old. Frontinus, a professional soldier who had been governor of Britain in A.D. 80, was a perfectionist in all the many matters to which he turned his hand. He once wrote, 'I consider it to be the first and most important thing to be done, as has always been one of my fundamental principles in other affairs, to learn thoroughly what it is that I have undertaken.' Frontinus built nine aqueducts with a total length of 263 miles, much of which was carried on arches or through tunnels. The Appian aqueduct ran underground for ten of its sixteen miles. By the year A.D. 100, the city of Rome was so well supplied with water that the daily consumption was estimated to be 35,000,000 cubic feet, which, reckoning the population of Rome at that time to be half a million, is equivalent to about 500 gallons per head per day—or over ten times the present domestic consumption of London which is 48 gallons per head per day.

Came the decline and the fall of the Roman Empire, its elaborate system of conduits, aqueducts and tunnels met the same fate as the rest of Roman building and engineering. What was not dismantled for use as building-stone fell into ruin and decay, and tunnelling became a lost art. The Romans had held dominion over practically all the civilised world for more than a thousand years—and over a substantial part of it for another thousand. But the thousand years that followed their departure was one of stagnation during which nothing of any consequence was achieved underground, and even at the end of this period it was to be several hundred years before the art of tunnelling was developed beyond the standard of the Romans.

3

The Beginnings of
Modern Tunnelling

THE TUNNELS BUILT during the millennium that followed the fall of the Western Empire were not public works and were almost entirely confined to crypts under castles, cloisters, monasteries and churches, or secret foot-passages connecting the castles of mediaeval barons. In 1450 an attempt was made to revive the grand-scale tunnelling of the Romans—when a tunnel was started that was intended to run under the 6,158-foot Alpine pass of Col di Tende, to connect Nice and Genoa. The scheme was dropped, probably through lack of knowledge and technique, but was subsequently continued in 1782—by which time tunnel engineering had again become highly developed. The work was finally abandoned in 1794 when about $1\frac{1}{2}$ miles had been excavated.

When the Moscow underground railway was built in 1935, an extensive and elaborate system of tunnels was uncovered that had been built by Ivan the Terrible (1530–1584), as a venue for some of his peculiar recreations, which included watching human beings being eaten up by wild animals. One passage extended for some distance and emerged at a place that, at the time of the tunnel's construction, had been a forest populated by wild beasts. This maze of underground excavations included courtyards, dungeons and catacombs as well as living rooms and an armoury. But examples of large-scale tunnelling before the seventeenth century are rare and the history of the modern era of tunnelling starts at the beginnings of the 'Canal Age'.

Canal tunnels

The builders of canals drove tunnels for exactly the same reason as the Romans: because water will not flow uphill. It is true that by means of a series of locks a canal can climb a hill and descend the other side, but a good supply of water is required at the summit especially on a busy waterway. Although, in theory, one lockful of water moves all the way down a flight or series of locks, thus serving *all* of them, in practice lock-gates leak and there is a steady loss of water in addition to that which is used in moving the traffic up and down. To make up this loss usually meant constructing large reservoirs or the installation of machinery to pump water up from streams on lower levels. In addition to these expenses, locks required regular maintenance and the services of a lock-keeper. Most inconvenient of all was the considerable loss of time involved in passing boats through locks. Then again the expense of building a flight of locks could be very great indeed and certainly an item to be taken into consideration when deciding whether to go over or around or *under* an obstructing piece of land.

The Languedoc canal, or Canal du Midi, connecting the Bay of Biscay with the Mediterranean Sea, was the first of the great European canals. Along its length of 148 miles it has 119 locks, and its summit is 620 feet above sea-level. It has the distinction of possessing the first canal tunnel ever built which, in its turn, was the first tunnel to be excavated by gun-powder, the use of which represented the greatest advance ever made in rock tunnelling. Last, but not least, it was the first tunnel constructed for transportation and thus was the 'father' of tunnels as we know them today. Situated at Malpas, about 6 miles from Béziers, the tunnel is 515 feet long, 22 feet wide and 27 feet high; after two years' work it was finished in 1681 and arched in some ten years later. No details of the story of the building of the Malpas tunnel have survived, but it must have been a trying and violent task. The problems involved in the use of explosive in the confined space of a tunnel and the behaviour of rock under such con-ditions had never been studied and it is unlikely that the work was carried out without considerable loss of life. Nevertheless it *was* carried out and the Malpas tunnel was the fore-runner of an era of canal-tunnel building that continued through almost 150 years, leading to the near frenzy of tunnel driving that accompanied the railway mania of the early nineteenth century.

James Brindley

The father of tunnelling in Britain was, undoubtedly, the self-taught engineer James Brindley (1716–72). A craftsman of more than ordinary brilliance, Brindley constructed no less than 360 miles of canals while remaining almost illiterate to the end of his life. Even his most immense projects were embarked upon and completed without the use of written calculations or drawings. The son of a crofter, he was born in Derbyshire. After serving an apprenticeship as a millwright and later practising as a wheelwright, he designed and built an engine for draining a coal-pit, in 1752. In 1755, he had built the machinery for a silk-mill at Congleton, Cheshire. Four years later he was engaged by the Duke of Bridgewater to dig a canal from the Duke's collieries at Worsley, near Manchester, into the city itself.

Throughout his canal-building career Brindley showed a marked aversion to locks, preferring, wherever possible, long flat stretches of dead water. Accordingly, after being asked to give an opinion on the best route for the proposed canal, Brindley, after making what he termed an 'ochilor servey or a ricconitoring', advised the Duke that instead of carrying the canal down to the River Irwell by a flight of locks and then up again the other side, it should be carried over the river and be constructed on one level throughout. Brindley's proposals were regarded by his fellow engineers as lunatic, for not only would they involve confining and carrying, over a mighty embankment, a mass of water within a watertight trunk of earth, but in addition a high bridge would have to be built across the Irwell to carry heavy ships over the heads of other ships below! The story of how that embankment and aqueduct were built is not for this book, but suffice to say that Brindley's plans were accepted in spite of the scorn of his professional colleagues and the bridge and earthworks can be seen today. At the Worsley end Brindley tunnelled the canal into a hill for a considerable distance, so that it connected with the colliery workings themselves. At this time the tunnel penetrated a mile into the rock. By 1878 it extended to a combined length of over 40 miles in all directions underground.

At the Manchester end of the canal, Brindley was equally daring. It had been originally intended to terminate the canal at the foot of Castle Hill and to hump the coal uphill in wheelbarrows. Instead the canal was again tunnelled into the hill and the coals were hoisted up a shaft by a crane, worked by a water-wheel, of 30 feet diameter, powered by the waterfall on the River Medlock. The Worsley canal with its terminal tunnels offers a good example of transport economics, for when it was opened in 1761 the cost of coal in Manchester dropped by half.

While still engaged on the Duke of Bridgewater's canal, Brindley was busy carrying out a much larger enterprise: the canal that connects the Mersey with the Trent, and both with the Severn. Then known as the Grand Trunk, the canal passed through five tunnels: the Harecastle, 2,880 yards long; the Hermitage, 130 yards; the Barnton, 560 yards; the Saltenford, 350 yards; and the Preston-on-Hill, 1,241 yards. All were 17 feet 4 inches high by 13 feet 6 inches wide, except the Harecastle which was only 9 feet high. When it was known that Brindley intended to cut through the Harecastle Ridge, his detractors saw the scheme as evidence of his insanity and the project became known as the 'Air Castle', but no sooner had the necessary Act been passed through Parliament than work was begun to pierce the ridge. Work started in 1766 and the cutting of the tunnel took no less than eleven years—an average progress of 2 feet per day! Excavation was effected in both directions from a number of shafts sunk to grade along the tunnel's path, the spoil being loaded into buckets that were drawn up the shafts by horse-gins.

Then, to deal with the small quantities of water that were encountered, pumps were installed powered by windmills and watermills. As the shafts penetrated the hill, water was met with in quantities sufficient to drown out the men at the bottom of the shafts and operations had to be postponed while a Newcomen engine was erected to keep the shafts dry. The presence of the water was, in fact, known to Brindley before it had manifested itself —the engineer relied on it to feed his canal at that point which was the summit level.

The Harecastle tunnel was finally completed in 1777—five years after Brindley's death. James Brindley was one of the most remarkable instances of self-taught genius in an era that abounded in such men. To the last he never lost his roughness of character and demeanour, but neither did he lose his almost uncanny grasp of mechanical situations. His brother-in-law once said of him, 'When any extraordinary difficulty occurred to Mr Brindley in the execution of his works, having little or no assistance from books or the labours of other men, his resources lay within himself. In order, therefore, to be quiet and uninterrupted while he was in search of the necessary expedients, he generally retired to his bed; and he has been known to be there one, two, or three days, till he had attained the object in view. He would then get up and execute his design without any drawing or model.'

'Legging'

The widespread building of canal tunnels brought into being a new and most unlikely 'profession', that of 'legging'. For the sake of economy,

tunnels were made as narrow as practicable and seldom contained tow-
paths, the horses being led over the hill while the boat was propelled
through the tunnel by 'leggers', who, lying on legging-boards which they
temporarily affixed to the side of the craft, 'walked' or 'legged' along the
side of the tunnel. The job was not without its dangers—for we find the
Grand Junction Canal Company advising leggers to 'strap themselves to a
short Cord fixed to the Boat to prevent their being drowned'. The same
Company later complained of 'the nuisance arising from the notoriously
bad characters who frequent the neighbourhood of the Tunnels upon the
plea of assisting Boats through them'. To put an end to this nuisance, the
Company registered and licensed its leggers and compelled them to wear
an official badge. Later, with the advent of the steam tug, this somewhat
precarious occupation became a thing of the past, although the risk of
falling off a legging-board and drowning inside the tunnel gave way to the
even greater risk of suffocation from smoke in those inadequately ventilated tunnels. In September 1861, the entire crews of two tugs, engaged in
towing a string of boats through the Blisworth tunnel, were overcome by
smoke, steam and what a contemporary report called 'decomposed air'.
When the tugs emerged two of the crew were dead, one having been
roasted by falling across the engine. What with drowning, roasting and
suffocating it is no wonder that many of the longer canal tunnels gained
the reputation of being haunted.

It is still said that the second Harecastle tunnel is haunted by a
phantom of hideous mien and unprecedented malignity, known locally as
'Kidgroo Boggart', while the Crick tunnel on the Leicester section of the
Grand Union is inhabited by the ghost of an amiable old woman who, if
she likes you as you pass through her domain, will come on board and
cook you a meal. Her name is Kit Crewbucket—the connection with the
Harecastle Horror being obvious.

It may be said that the apparent ease with which Brindley pierced any
obstacle that stood in his path inspired the canal builders of Europe to
follow his example, for the next half-century saw a wave of tunnelling on
a substantial scale. Great as were the achievements of the eighteenth-
century tunnel engineers, all their works had been driven through hard
ground and by the end of the century no one had yet succeeded in tunnelling under a river. The advantages of sub-aqueous passage-ways were
fully appreciated both by engineers and promoters and many a plan was
laid to burrow beneath rivers. All were shelved upon examination, until in
1807 the great Richard Trevithick was engaged by a private promoter to
build a tunnel for foot-travellers under the River Thames between
Rotherhithe and Limehouse.

Richard Trevithick

Like Brindley, Richard Trevithick (1771–1833) was a self-taught engineer. An expert on mine-pumping, he invented a number of devices for the improvement of deep mining and designed and built a high-pressure non-condensing steam-engine, which rivalled the low-pressure steam-vacuum engine of Watt. In 1801 he demonstrated his steam road-locomotive by carrying the first load of passengers ever to be conveyed by steam. All this and much more had been achieved by Richard Trevithick when, at the age of thirty-six, he was given the job of building the world's first underwater tunnel. 'Last Monday I closed with the tunnel gents,' he wrote in August, 1807. 'I have agreed . . . to receive £500 when the drift is half way through, and £500 when it is holed on the opposite side . . . this will be making £1,000 very easily and without any risk of loss on my side.'

It was intended first to drive an advance, or pilot, tunnel, 5 feet high, 3 feet wide at the bottom and 2 feet wide at the top, and, when complete, to enlarge it to a full size of 16 feet high and 16 feet wide. This pilot tunnel was to be timbered to keep the loose earth from falling in and a steam-engine was set up on the surface to keep it dry, and to power a ventilating fan. Despite the fact that the heading would only allow one man to dig at one time and that he could not even stand upright, the tunnel made good progress and in the space of two months Trevithick had burrowed 400 feet. The small quantities of water that were encountered were easily dealt with by the pumps.

It is difficult to imagine what working conditions were like in that black hole as it crept on its way beneath the bed of the river. With malodorous water continually soaking their clothes, and breathing foul air as they worked by the guttering light of candles, Trevithick and his men were again and again half buried by quicksand, 'fine as flour', that broke in from the roof or from the work-face. Sometimes rock was encountered, to be laboriously penetrated with hammer and chisel, for it would not have done to use explosive in such a confined, unstable situation. Beds of oyster shells and fossils of trees far older than the Thames itself were encountered, together with sodden clay and compressed gravel that seemed to indicate that the river bed could not be far above. To ensure that the soft mud of the river was never less than 30 feet above him, Trevithick took borings by means of hollow iron pipes which he thrust through the roof until water flowed through them.

After a year of this work Trevithick reported that his heading was 950 feet long and that only another 140 feet separated him from the north bank of the river. A few days later, in the spring of 1808, disaster struck. Trevithick was in the tunnel supervising the work of the man at the face while behind them, and of necessity in single file, were three men erecting the timbering. Suddenly and without warning the sandy face of the heading fell in and before this disastrous event even had time to register with the engineer the water had rushed in. The resulting current of air blew out the candles as the five men turned to run. But run they could not as they ducked under the low timbers of the roof and stumbled over the débris and loose sand on the floor. Trevithick, himself over 6 feet tall, was bent almost double as, fourth in line, he lurched with his workmen towards the ladder running up the tunnel shaft. When the ladder was reached there was a delay as each man mounted it and the last up ascended the shaft with the water at the level of his neck—a level that he was just able to maintain until he reached the surface. Within a week, Trevithick had located the hole in the river-bed and had filled it with bags of clay dropped from boats. Gravel was then heaped over the clay bags to make a solid seal and the tunnel was pumped out. It is a tribute to the engineers' timbering work that after such a violent inundation there was a tunnel still to pump out. The work of stopping the leak and drying out must have been done in very good time, for six days after the flooding the miners were back at work and the heading was beyond low water mark on the north shore.

A month later the Thames again penetrated the narrow passage-way, the miners were again forced to abandon the workings and this time for good. The Thames Water Bailiff refused Trevithick permission to dump any more bags of clay into the river as, he said, they constituted an impediment to shipping. Work on the flooded heading was stopped, the works abandoned and sealed off. Accepting their failure the directors of the company announced: 'Though we cannot presume to set limits to the ingenuity of other men, we must confess that, under the circumstances which have been so clearly represented to us, we consider that an underground tunnel, which would be useful to the public . . . is impractical.'

Trevithick countered this by proposing to sink a series of coffer-dams along the tunnel line on the river-bed, to excavate a ditch within them and to bury a cast-iron, sectional pipeline. The scheme was turned down as impossible, and work on the tunnel was finally abandoned. (In the present century, American engineers have spanned the beds of several rivers by this very method.) It was to be another thirty-four years before a tunnel was driven under the Thames, and that was still to be the first underwater tunnel in the world.[1]

---

[1] There are records of a footpath under the Euphrates built by Queen Semiramis in 2180 B.C., but its existence is by no means certain. Even if it was built it was not a tunnel in the true sense of the word, as it was constructed by the 'cut and cover' method after the river had been diverted during the dry season. It is doubtful even if Queen Semiramis ever existed.

Plate 1. The breakthrough of the St Gotthard tunnel, at 11.00 A.M., 29 February 1880.
An artist's impression, and presumably somewhat over-romanticised.

Plate 2. Louis Favre, the contractor for
the St Gotthard tunnel.

Plate 3. A group of working men in front of the Simplon tunnel, 1899.

Plate 4. Working inside the Simplon tunnel.

Plate 5. Working by hand on the face of the Simplon tunnel.

Plate 6. Inside the great Loetschberg tunnel, $9\frac{3}{4}$ miles long.

Plate 7. Three divisions of Brunel's shield for the Thames tunnel.

Plate 8. The method of removing soil from behind the shield for the Thames tunnel.

4

The Brunels: Father and Son

TO MARC ISAMBARD BRUNEL (1769–1849), the French engineer who emigrated to this country in 1793, go three distinctions: he was the father of Isambard Kingdom Brunel (1806–1859), the greatest civil engineer in history; he invented the first tunnelling-shield and thus revolutionised the practice of tunnelling; and he drove the first tunnel under a navigable waterway. These eminent achievements were unified when Marc, with his son as resident engineer, used the shield to drive a carriageway under the River Thames between Rotherhithe and Wapping.

Marc Brunel had settled in England during the time that Trevithick was toiling in the mud under the Thames, and he took a keen interest in the project. The inundations of water that finally led to the abandonment of the work led him to the inescapable conclusion that, as in all soft-soil tunnelling, the top and sides of the excavation had to be supported until the brickwork lining was built and, furthermore, tunnelling beneath the Thames would for some of the way be virtually tunnelling through mud; it would therefore be necessary to hold up the face of the heading as the miners advanced. The principle of a tunnelling-shield came to Marc Brunel when he was studying the tunnelling action of the 'ship-worm', *Terado navalis*, which was the cause of much damage to the wooden ships of the day. Brunel's first tunnelling-shield, patented by him in 1818 but never built, consisted of a 12-foot auger-blade encased in an iron cylinder. The blade was to be rotated manually by a miner, while at the same time the cylinder was to be propelled along the line of the tunnel by jacks working against a brickwork lining which was to be built as the tunnel progressed.

In practice it would have been beyond the strength of even a large team of men to turn the cutter with sufficient force to slice through a 12-foot diameter circle of mud, let alone one mixed with the clay and gravel that would be met, and it is surprising that an engineer with Marc Brunel's experience could not appreciate this. The important thing was that the principle was correct—as the engineer himself expressed it 'an ambulating coffer-dam, travelling horizontally'. Brunel's second design dispensed with the auger-blades and set the pattern for the principle of shield-tunnelling for over a century. When the promoters of the still-existing Thames Tunnel Company learned of the invention they studied the design closely and had long discussions with the engineer. The upshot of this was that at a meeting of businessmen at the City of London Tavern on 18 February 1824, a large sum of money was subscribed to use the shield to build their belated tunnel under the River Thames. In the following July, Marc Brunel was appointed engineer-in-charge at a salary of £1,000 a year with an additional £5,000 for the use of his shield. He was to receive another £5,000 when the tunnel was complete.

The first underwater tunnel—the Thames tunnel

Work began on the world's first sub-aqueous tunnel on 2 March 1825 at Cow Court, about three-quarters of a mile due west of Trevithick's starting-point near St Mary's, Rotherhithe. A large brick cylinder was first built above ground on the site of the intended shaft. The cylinder, 50 feet in diameter and 42 feet high, was securely braced by iron tie-rods, secured to cast-iron rings at top and bottom. The plan was to dig away the earth from inside and beneath the cylinder so that it would sink into the ground by its own weight. The 910-ton brick shaft was built in three weeks before a constant crowd of tourists and sightseers; then on April 21 the pick and shovel gangs climbed inside and began sinking it. A fourteen horse-power engine performed the dual task of keeping the workings dry and powering an endless chain of buckets that lifted the muck out of the cylinder. Next to the shaft, on the outside, a well was dug, keeping always a few feet deeper than its big neighbour, to indicate to the engineer what sort of soil and what obstacles his main shaft would meet.

The cylinder sank into the earth and on June 6 it was at full depth. The height of the shaft was then further increased by 20 feet, this being done by the more conventional method of underpinning. At the bottom of the shaft a large reservoir was sunk to drain the workings and this in turn was emptied by steam-driven plunger pumps. With the shaft thus completed the shield was installed and the tunnel itself started. This was to be a rectangular mass of brickwork, 37½ feet wide and 22 feet high, forming two parallel horseshoe-shaped arches, each 14 feet wide and 17 feet high. Marc Brunel's 'ambulating coffer-dam' consisted of twelve massive frames made of cast- and wrought-iron, each nearly 22 feet high and a little over 3 feet wide; when placed side by side, like books on a shelf, against the face of the excavation, they formed a shield with top, bottom and sides, holding up the earth 9 feet in advance of the brickwork. The frames were divided into three storeys or compartments, each about 7 feet high by 3 feet wide, and occupied by a single miner. Thus the assembled shield accommodated thirty-six men, all occupying separate working chambers. Each frame supported a series of boards called poling-boards which, by means of screw-jacks, two to each board and abutted against the frames, were pressed against the earth at the face of the drift. Arranged horizontally, these 500 boards were each 3 feet long, 6 inches wide and 3 inches thick.

The feet of the frames were broad iron shoes connected together by ball joints, while the roof of the uppermost cell was a pivoted plate called a stave that corresponded to the shoe at the bottom. The earth at the sides was kept up by staves fixed to the outermost frames. The operation of tunnelling was carried out by each man removing one of his poling-boards and excavating the earth behind it to a depth of about $4\frac{1}{2}$ inches. He then replaced the board and drove it forward by means of the screw-jack. This process was repeated until all the 500 boards had been moved forward. The frames were then advanced in the following way. The foot of one of the frames was lifted up by means of its jointed leg and advanced $4\frac{1}{2}$ inches when it was then pressed back on the ground. Then the other foot was similarly treated and the frame itself pushed forward by means of large screw-jacks abutted against the brickwork of the arch which was advancing just behind it. As the large abutting screws would not have been able to move the frames while the poling-boards were in position, the butts of these were shifted sideways at each movement of the frame so that they pressed not against the frame to which they belonged but to the one alongside. When the frame had been moved forward it received back its own poling-boards, plus those belonging to its neighbours, so that they in turn could move forward. The bricklayers worked so close to the advancing frames that the brickwork was carried right up to the staves which formed the shield around the working frames. With the whole excavation thus supported and protected, the only exposed areas of ground were those left by the removal of a poling-board.

The shield began its slow journey on 25 November 1825, and it was apparent from the start that the geologist's report on the nature of the ground to be encountered had been far from accurate. With the shield a good 100 feet from the river's edge, water began pouring into the workings through faults in the clay. Marc Brunel had intended to cut a driftway below the tunnel to carry off any water that might be encountered, but this proposal had been vetoed by the Company on account of cost. The directors were to regret that decision, for apart from the subsequent disasters that it caused, forty men were constantly employed in the workings, operating bucket pumps to keep the shield clear of water. Yet even so, the men in the lower compartments spent much of their time knee-deep.

Early conditions underground

Working conditions in mine or tunnel in the early nineteenth century, even at their best, would be unbearable by modern standards, but in the tunnel under the Thames they were at their worst. Working by candle and lantern in that unventilated passage, the men at the face were continually saturated with spurts of evil-smelling water or liquid mud from a river that was no more than an open sewer. The very air they breathed was heavy with corruption and it was no wonder that the sick-list was always long. Gastric infections and diseases caused through constant damp were commonplace, but worst of all was 'tunnel sickness', an often fatal affliction which at its best, or worst, left men permanently blind. In these conditions the men worked a two-shift system, eight hours on and eight off, around the clock, six days a week. The Brunels, father and son, worked even longer hours, one of them always being present in the workings and frequently both. Even when resting, Marc Brunel insisted on being woken every two hours to learn the condition of the ground. As engineer in charge, Isambard Kingdom Brunel was little more than a boy, but as L. T. C. Rolt points out in *Isambard Kingdom Brunel*, 'The fact that he was barely twenty years of age occasioned little remark; yet today, when an educational fetish prolongs childish irresponsibility far into adolescence, it seems almost incredible that such an immense burden of responsibility should have been laid on such young shoulders and that it should have been carried with such distinction.'

Marc Brunel had estimated an average rate of travel of 3 feet per twenty-four hours but on 11 May 1826 he was recording the completion of only the first 100 feet of tunnel. Thereafter progress speeded up to an average of 1 foot in twenty-four hours. The following entry in Marc Brunel's diary describes a typical day in the workings.

> September 8 1826.—About 2 p.m. I was informed by Munday that water was running down over No. 9 [the frames were designated by numbers]. I went immediately to it. The ground being open, and consequently unsupported, it soon became soft, and settled on the back of the staves, moving down in a stream of diluted silt, which is the most dangerous substance we have to contend with. Some oakum was forced through the joints

of the staves, and the water was partly checked. Isambard was the whole night until three in the frames. At three I relieved him. He went to rest for about a couple of hours; I took some rest on the stage.

September 9.—Towards noon the stream changed its character. The clay, being loosened by the water, began to run, but it thickened gradually. It was late in the evening before the loosened clay acquired the consistency of a loose pudding, which covered the staves, and made them a complete shield against further irruption, or rather, oozings of mud . . .

September 12.—The water, bringing with it a sort of clay broken in small particles, increased to an alarming degree. In consequence of this continued displacement of the silt and clay, a cavity had been formed above the staves. At about three . . . the ground fell upon the staves with great violence . . . Isambard was at that moment in the upper frames, and he gave directions for increasing the means of security. During the night in particular things presented a very unfavourable appearance. The men, however, were as calm as if there were no other danger to be dreaded than wet clothes or the splashing of mud. I observed the men in the lower cells were *sound asleep.*

September 13. . . . Isambard has not quitted the frames but to lay down now and then on the stage. I have prevailed on him to go to his bed . . . but he has not since last Friday (the 8th).

December 20.—An accident of an alarming nature occurred. The poling-screws of Nos. 10 and 12, being on No. 11, Moul, the miner in that frame, removed his butting screw; the consequence was that the frame started back, the polings and poling-screws fell down with a tremendous crash, and the ground followed to a considerable extent. This is the most formidable accident that has yet occurred in the face of the work. The ground was fortunately unusually firm, and no fatal consequences ensued.

But in spite of mishaps, accidents and delays, the two Brunels made slow but steady progress in their subterranean struggle with the elements of earth and water. As if to remind them of the nearness of the river above their heads, various articles were found among the muck that was excavated from the work-face—nails, pieces of china, a buckle and items of ships' rigging, but still the dreaded disaster did not happen. 'January 4, 1827—Every morning I say, "Another day of danger over!"'

On April 29, when Isambard was at breakfast in the tunnel, a workman ran back to him from the frames, ashen-faced, shouting, 'It's all over, it's all over, the river's in and they're all drowned except one.' Isambard and his assistant, Gravatt, dashed from the breakfast table to the work-face, where they observed that a small lump of clay had fallen from the top of the shield.

On 13 May 1827, Marc Brunel wrote, 'So far the shield has triumphed over immense obstacles, *and it will carry the tunnel through.* During the preceding night the whole of the ground over our heads must have been in movement, and that too at high water. The shield must have therefore supported upwards of *six hundred tons*: it has walked for many weeks with that weight twice a day over its head.'

On the same day he also wrote: '. . . a disaster may still occur. *May it not be when the arch is full of visitors!* It is too awful to think of it.'

Five days later the arch was full of visitors and Marc records that he attended Lady Raffles '. . . most uneasy all the while, as if I had a pre-sentiment'. During the evening while the visitors were still in the tunnel an assistant-engineer, Richard Beamish, had just relieved Isambard in the frames when he noticed that No. 11, which was ready to be worked forward, was making more water than usual.

A little later there was a cry for help from No. 11, but before Beamish could get into the frame a mighty torrent of water swept out of it, taking one of the miners with it. Beamish made a desperate effort to breast the water and get into the frame until one of the men from No. 9 caught him by the arm shouting: 'Come away, sir, come away; 'tis no use, water's rising fast.' The men dashed for the spiral stairway of the shaft with the water roaring in behind them and plucking up timber and stores as it went. Half-way between the shield and the shaft there was a wooden office building; as this was overwhelmed by the wall of water the heading was plunged into darkness. The men reached the shaft without time to spare and there they found Isambard hastening others up the stairs. William Gravatt was in the tunnel at the time and as his account was taken down in shorthand immediately after the inundation it is worth quoting in full.

I was above with I. Brunel looking over some prints, Beamish being on duty. Some men came running up and said to Isambard something I did not hear. He immediately ran towards the works, and down the men's stair-case. I ran towards it, but could not get down. I leaped over the fence, and rushed down the visitors' stairs, and met the men coming up, and a lady, who I think was fainting. Met Flyn on the landing place, who said it was all over. I pushed on, calling him a coward, and got down as far as the visitors' barrier. Saw Mr Beamish pulled from it. He came on towards the shaft walking. I went up to ask him what was the matter. He said it was no use resisting. The miners were all upon the staircase; Brunel and I called them to come back. Lane [the foreman bricklayer] was upon the stairs, and he said it was of no use to call the men back. We stayed some time below on the stairs, looking where the water was coming in most magnificently.[1] We could see the farthest light in the west arch. The water came upon us so slowly that I walked backwards speaking to Brunel several times. Presently I saw the water pouring in from the east to the west arch through the cross

[1] Gravatt was not the only one present to be affected by the 'magnificence' of the scene, for Beamish recorded that 'the effect was splendid beyond description; the water as it rose became more and more vivid from the reflected lights of the gas'.

arches. I then ran and got up the stairs with Brunel and Beamish, who were then five or six steps up. It was then we heard a tremendous burst. The cabin had burst, and all the lights went out at once. There was a noise at the staircase, and presently the water carried away the lower flight of stairs. Brunel looked towards the men, who were lining the staircase and galleries of the shaft, gazing at the spectacle, and said, 'Carry on, carry on, as fast as you can!' Upon which they ascended pretty fast. I went up to the top and saw the shaft filling. I looked about and saw a man in the water like a rat. He got hold of a bar, but I afterwards saw he was quite spent. I was looking how to get down, when I saw Brunel descending by rope to his assistance. I got hold of one of the iron ties, and slid down into the water hand over hand with a small rope, and tried to make it fast round his middle, whilst Brunel was doing the same. Having done it he called out, 'Haul up.' The man was hauled up. I swam about to see where to land. The shaft was full of casks. Brunel had been swimming too.

On the morning of the day following the disaster—a Sunday—the curate of Rotherhithe church referred to the incident in his sermon as '. . . but a just judgement upon the presumptuous aspirations of mortal men', but that did not prevent Brunel the younger from later that same day borrowing a diving bell from the West India Dock Company and descending to the river-bed to survey the damage. Here he found a deep hole that was the result of gravel dredging and protruding from the bottom of the hole was the top stave of one of the shield frames. Brunel was actually able to stand with one foot on the footboard of his diving bell and the other on the shield.

Describing his impressions of this visit to the bed of the Thames, he wrote, 'What a dream it now appears to me! Going down in the diving bell, finding and examining the hole! Standing on the corner of No. 12! The novelty of the thing, the excitement of the occasional risk attending our submarine excursions, the crowds of boats to witness our works all amused.'

The following day a large mass of bags full of clay and hazel rods was lowered into the hole and on top of these a raft loaded with 150 tons of clay was sunk. Then the pumps in the shaft were set to work and the water level dropped. The following day it rose to its original level and Isambard, again descending to the river-bed, found that the tide had tilted the raft and left an opening on its west side. The raft was refloated and towed ashore and the breach was finally filled by laying iron bars across the hole to form a bed for a further load of clay-filled bags.

On July 11, with 1,950 cubic feet of clay stopping up the hole in the river, the pumps again started work and by the 25th the shaft and 150 feet of tunnel were dry again. On the 27th Brunel, accompanied by a picked team of his men, explored the rest of the tunnel by boat which, towards the end, they had to propel by pushing with their hands on the roof. Eventually the boat came to rest against a pile of silt over which they crawled until they reached the frames. These were found to be intact and undisturbed. Brunel later wrote of

> . . . the low, dark, gloomy, cold arch; the heap of earth almost up to the crown, hiding the frames and rendering it quite uncertain what state they were in and what might happen; the hollow rushing of water; the total darkness of all around rendered distinct by the glimmering light of a candle or two, carried by ourselves; crawling along the bank of earth, a dark recess at the end—quite dark—water rushing from it in such quantities as to render it uncertain whether the ground was secure; at last reaching the frames—choked up to the middle rail of the top box—frames evidently leaning backwards and sideways considerably—staves in curious directions, bags and chisel rods protruding in all directions; reaching No. 12, the bags apparently without support and swelling into the frame threaten every minute to close inside brickwork. All bags—a cavern, *huge, mis-shapen* with water—a cataract coming from it—candles going out . . .

On 26 July 1827, two months and eight days after the disaster, the tunnel was clear of water and the great shield continued its ponderous progress through the mud. By the last day of the year, 600 feet of the tunnel had been completed, but on January 12 the water made another breakthrough which put a stop to the works for a much longer period. On this occasion the younger Brunel was working at the shield with two miners when the exposed 6 inches of ground suddenly swelled out towards them and water began gushing in. No. 1 frame then seemed to disintegrate before their eyes as the mighty torrent burst upon them, sweeping them like so much flotsam out of the cell and into the arch behind. In the instant darkness that followed the men struggled through the tunnel towards the shaft. Finding the spiral stairway blocked with a mass of escaping men, Brunel made for the west arch where there was another flight of stairs. Here a huge wave caught him, carrying him to the top of the shaft and safety. The two miners who were washed out of the frame with him were drowned, together with four others who were swept off the stairway by the same wave that saved Brunel's life.

The tunnel had again run into a cavity in the river-bed and this time 4,500 tons of clay were needed to fill the hole before the heading could be cleared. By the time this was done the Company had run out of money and although there was great public enthusiasm for a resumption of work it was not accompanied by cash subscriptions. The tunnel was bricked in and abandoned for the next seven years.

It was not until 1835 that the Thames Tunnel Company, with the help of a government loan, was able to resume work and by then Marc Brunel had made considerable improvements in the design of his shield. The new design included slings to connect the frames so that each frame could, when necessary, support its neighbour. There were also devices to facilitate changes of direction.

Five years after the recommencement of work with the new shield, the head of the tunnel neared the Wapping side of the river and in October 1840 the shaft on that side was begun. It differed from the Rotherhithe shaft in that it was slightly conical in shape to reduce friction during sinking. It was sunk to its full depth of 70 feet without underpinning. In March 1843 the tunnel was completed and opened to the public at a toll of a penny a head.

The Thames tunnel soon became one of London's biggest tourist attractions and the Company was quick to cash in on its popularity. Stalls were built along both its walls and in the cross-cuts that connect the two passage-ways, and these were rented out to sellers of souvenirs, cheap jewellery, cakes and ginger-beer. Then there were exhibitions of paintings, tight-rope walking, puppets and conjurers. But for all this the tunnel was a financial failure. In 1865 the near-bankrupt Company sold out to the East London Railway. The money they received went to repay the balance of a government loan and the businessmen who had invested over £180,000 in the project lost all their money.

The Thames tunnel is now part of the London underground railway system. The traveller between Wapping and Rotherhithe can still climb up and down the brick shafts and, while waiting for his train, he can imagine how the twin-arched passage-ways appeared when the tunnel was one of the world's wonders—ablaze with gaslight and gay with stalls and sideshows. A man standing in the middle of the tunnel, at night, when the trains are no longer running, may occasionally hear a thumping that reverberates along its length. This will be the engine of a motor-vessel passing along the Thames a few feet above his head.

The Wapping–Rotherhithe tunnel was over eighteen years in the building—a period of time that saw many changes in the world's communications. Indeed, when Marc Brunel was drawing up his plans in 1825, a passenger railway line had already been opened in England. This was the Stockton and Darlington line and it employed a novel method of transport: the steam locomotive.

5

Early Railway Tunnels

WRITING IN 1800 on the subject of canals, the engineer Thomas Telford (1757–1834) observed:

Since the year 1797 . . . another mode of conveyance has frequently been adopted in this county [Shropshire] to a considerable extent; I mean that of forming roads with iron rails laid along them, upon which the articles are conveyed on waggons containing from six to thirty cwt.: experience has now convinced us that in rugged country or where there is difficulty to obtain lockage, or where the weight of the articles of produce is great in comparison with its bulk, or where they are mostly to be conveyed from a higher to a lower level, in these cases iron railways are in general preferable to canal navigation.

While it is true that even the early locomotives could cope with inclines to some extent, the need for level tracks, where possible, was obvious. When faced with the choice of running round a hill or through it, the railway companies had two reasons for taking the latter course. Firstly the purchase and maintenance of land; secondly, and more important, the times of running their trains. A 5-mile detour hardly affected the cost of running a horse-drawn barge but it could add a great deal to the cost of running a coal-consuming engine, and on a busy line a tunnel could save a very considerable sum indeed. The builders of railways picked up where the canal builders left off, and as the railway revolution got into its stride the technique of tunnel construction made more progress in a few decades than it had in the preceding sixteen centuries. Then, as the railway system spread, there was hardly a town or village that was more than a dozen or so miles from a tunnel; even though commonplace many of them were major achievements of engineering.

The Liverpool and Manchester railway tunnel

The very first railway tunnel was begun in 1826. This was the Terrenoir (Black Earth) tunnel, on the single-track horse-drawn railway between Roanne and Andrezieux, in France. In the same year work started on the first steam-railway tunnel. This was the Wapping tunnel running under Liverpool between Liverpool Edge Hill and Park Lane goods-stations. The designer and Chief Engineer to the project was George Stephenson (who was yet to triumph with the 'Rocket' in the famous locomotive trials of 1829), with Charles Vignoles as resident engineer. The first shaft was sunk to grade in September and from its bottom a pilot heading was driven out. Early in 1827 it was found that the heading was causing serious subsidence to the foundations of some houses in Great George Square, owing to an error in direction. The result of this was that relations between Stephenson and Vignoles became strained and the latter resigned from his post. His place was taken by Joseph Locke, a pupil of Stephenson's, who carried out a thorough check on the original survey; this showed that the shafts had all been sunk several feet out of line and that the pilot tunnel was so much off course that the two headings would have missed each other. The tunnel is 22 feet wide by 16 feet high, the sides being perpendicular for 5 feet with a semicircular arch. Its total length of 2,250 yards is cut through red rock, blue shale and clay, but mostly through rock of every degree of hardness. Some of the shale encountered was so wet that great ingenuity was required to support the roof until the masonry lining was built. The conditions under which much of the work was done was described by a reporter who visited the tunnel.

> The extraordinary and stupendous undertaking of excavating a wide and lofty tunnel, from the shore of the Mersey, under the town, to the other side of Edge-Hill, for the passage of carriages to the line of open railway, is proceeding with as much celerity as the nature of the work will permit. The excavation began at different points on the line of the intended tunnel, the principal eyes being—one in White-street, one at the top of Duke-street, and one in Mosslake-fields; and some others are in a state of forwardness. They are each provided with the usual mining machinery for the hoisting up of the loosened material, and the tunnel being driven east and west from each eye, the miners will meet each other half way between the pits. The greatest progress has been made in that part of the tunnel running from the eye in Mosslake-fields; and where the surface of the intended railway appears to be about sixty feet beneath the ground. The substance met with in thus boring 'the bowels of the harmless earth' is almost entirely a reddish

free-stone, which forms the bed upon which the town, and most of the land in the immediate neighbourhood rests; and this circumstance is no doubt favourable, as the tunnel will not require archwork of brick to be thrown over it, except where the stone is unsound, or sand or other loose material is met with. A considerable quantity of the stone is regularly quarried for building. The excavation at Mosslake-fields has proceeded thirty or forty yards each way from the eye. It is about twenty-two feet in width, and sixteen in height, and where the stone above appeared somewhat insecure, it is merely supported by slender wooden props. The visitor may descend the eye in one of the buckets, with perfect security, and it is a novel and interesting sight to those who have never seen mining in its grander operations, to take a view of the noisy operations going on below, the echo of which is confined to the subterranean passage. Though numerous candles are burnt by the workmen, the 'darkness' of the cavern is but 'made visible', and the sound of the busy hammer, and chisel, and pick-axe, the rumbling of the loaded waggons along the railway leading from the further ends of the cavern to the pit, and the frequent blasting of the rock, mingling with the hoarse-sounding voices of the miners, whose sombre figures are scarcely distinguishable, form an interesting *tout ensemble* of human daring, industry, and ingenuity. The excavation at the top of Duke-street, as it is only a few feet under the lowest level of the deep quarry there, is of more desirable access to those who are timid, as it may be entered on foot by a small inlet below. Here the miners have also proceeded a considerable way, and also from the shaft in White-street, where some of the stone excavated is white, and of good quality for building purposes. The air in these caverns is as yet suitable for easy respiration, but we understand when the miners penetrate towards the points of junction, it will become unwholesome and confined. When one shaft has once a communication with another, the whole will be well ventilated. No water to impede the work has yet been found; and as the whole will be on a declivity when finished, the tunnel will be perfectly dry, and to all appearance may even be whitewashed. Ledges, or shelves, are left on the perpendicular sides of the rock, to form the abutments for the arch-work to be thrown over such parts as may require it. The several eyes are situated about twenty feet to the south of the tunnel, whence they communicate with it by an excavation running into it at a right angle. This was owing to an alteration of the line after the eyes were sunk. The small waggons used in conveying the stone and sand from the miners, to the bottom of the pit, are easily propelled by workmen along railways, so laid, that even in the dark they cannot diverge from the proper tract. We shall from time to time notice the further progress of this interesting undertaking, upon which we hope the proprietors will continue to employ as many men as the nature of the work will permit.

The London–Birmingham line

With the completion of the Liverpool and Manchester railway, George Stephenson turned his attention to the building of the London–

Birmingham line which, over its 112½ miles, passes through eight major tunnels with a total length of 7,336 yards. All were built between 1834 and 1838 and three of them are major accomplishments.

North-west of London the line passes under Primrose Hill for a distance of 1,164 yards. The ground here is London clay and was found by the tunnellers to be compact and dry—so dry that it was more difficult to work than stone. Entirely free from moisture when encountered, its absorbing qualities were so great that exposure to the air caused it to swell rapidly. The resulting pressure of the expanding clay upon the brick lining of the tunnel was enormous, and for days after the bricks were laid the lining creaked and cracked as minute chips of brick flew off. So difficult was it found to work in such clay that after a few months the contractors asked to be relieved of their obligation and the railway company finished the job with direct labour. The total cost of the Primrose Hill tunnel was over £280,000—a figure more than double Stephenson's original estimate.

The proposal to build the London–Birmingham railway was, understandably enough, the cause of much opposition from landowners through whose property the line was to run. The most violent in their opposition, and the most effective, were the Earls of Essex and Clarendon, who successfully debarred the Company from crossing their land. The result was a detour in the route of the line and a tunnel through the chalk of the Chiltern Hills 2 miles north of Watford Junction. Over a mile long, this tunnel cost £140,000 and ten lives. The ground to be pierced was of a treacherous and unpredictable nature, for the chalk contained many large fissures that were full of wet sand and gravel. In 1836 a gang of ten men were working in the centre of the tunnel when suddenly, according to one contemporary account,

> the whole mass of soil . . . gave way, completely burying ten men who were at work below. They were engaged in fixing one of the iron rings which are built into the top of the tunnel to support the brickwork of the shafts, and from all that could be learnt from observation—for not one was spared to tell the tale—it appeared one of the men had cut away some of the chalk to obtain more room to fix ironwork, and by so doing had penetrated so near the gravel that it broke in in an instant, and entirely filled up the space leaving them not a moment's time to save themselves.

It was also the successful opposition to the route of the London–Birmingham railway that led to the building of the tunnel under Kilsby Ridge near Rugby—an undertaking that ranks as one of the epics of nineteenth-century engineering. The opposition in this instance came from the people of Northampton and it resulted in taking the railway farther west than was originally intended and driving a tunnel through the ridge that now barred the path of the line. The tunnel is 2,400 yards long and runs at an average depth of 160 feet below the surface.

Trial borings along the proposed line indicated shale, a kind of soft slate. Work had scarcely begun when it was discovered that between two of the trial shafts at 200 yards from the south end of the tunnel there was an extensive quicksand under the clay that had, against long odds, escaped preliminary probes. No less than eighteen shafts had been sunk and the tunnel was being driven from the bottoms of these as well as from the two ends when, without warning, the roof of one of the headings gave way and a deluge of water burst in. The men working in the tunnel at the time only escaped by means of a makeshift raft which they 'legged' to the lower end of the shaft, up which they were hoisted by ropes. Again the contractors gave up and George and Robert Stephenson took charge of the work. Pumping engines were installed in an attempt to drain the flooded section but they were unable to master the level—at times the water continued rising in the shaft when the pumps were working full out. Eventually it was decided that specially large and powerful steam pumps were required to control the water, and these were ordered to be built as soon as possible. In the meantime, on Robert's suggestion, a drift was run along the heading from the south end of the tunnel with a view to draining off the water. When the drift was near completion it ran into quicksand which completely filled it—fortunately without loss of life. There was now no alternative but to await the delivery of the new pumps. When these were installed they were worked day and night for eight months, drawing up the water at an average rate of 2,000 gallons a minute. Even this massive effort did not suffice and it was found necessary to sink seven more shafts and install them all with more pumping engines before the water could be mastered.

When the work was at its height, no fewer than 1,250 men and 200 horses were employed at Kilsby Ridge. The tunnel was completed in June 1838 at a cost of £300,000 or £100 per lineal yard.

The driving of the eight tunnels on the London–Birmingham railway was, considering the difficulties encountered, a great achievement carried out within a remarkably short space of time. And yet it was only a small part of the whole work, the magnitude of which was without precedent in engineering history. Samuel Smiles compared the building of this railway with one of the greatest feats of ancient engineering—the Great Pyramid of Egypt. The pyramid occupied the labour of 300,000 men for twenty years and it has been estimated that the energy expended on it was equivalent to lifting 15,733,000,000 cubic feet of stone to a height of 1 foot. The labour spent in constructing the London–Birmingham, if reduced in the same way, amounted to 25,000,000,000 cubic feet *more* than was lifted for the Great Pyramid while the work was done by 20,000 men in less than five years. Furthermore, as Samuel Smiles again pointed out in *Lives of the Engineers*, 'whilst the Egyptian work was executed by a powerful monarch concentrating upon it the labour and capital of a great nation, the English railway was constructed, in the face of every conceivable obstruction and difficulty, by a company of private individuals out of their own resources, without the aid of Government or the contribution of one farthing of public money'.

Smiles' striking comparison, it must be remembered, refers only to the building of the 112½ miles of railway between London and Birmingham. The twenty-five years that followed the opening of the first passenger railway (the Stockton–Darlington), in 1825, saw the construction in the British Isles alone of no less than 6,890 miles of line and the magnitude of the entire work of railway building in Britain can be appreciated from the following table.

Railways of Great Britain and Ireland

Year	Capital paid up	Miles opened	Net receipts (per annum)
1854	£286,068,794	8,054	£11,009,519
1860	£348,130,127	10,433	£14,579,254
1865	£455,478,143	13,289	£18,602,582
1870	£529,908,673	15,537	£23,362,618
1875	£630,223,494	16,658	£28,016,272
1877	£674,059,048	17,077	£29,155,350
1879	£717,003,469	17,696	£29,731,430
1880	£728,316,848	17,933	£31,890,501
1887	£845,971,654	19,578	£33,880,110

The railway 'navvies'

The work force that carried out this mighty and unprecedented undertaking, that built the embankments and excavated the cuttings and tunnels was a unique nomadic class born of the canal era and matured by the railway age. Extremely well paid, the 'railway navvies' wandered from one project to another, living in shanty towns that were slung together for their benefit and moved with the work. Their clothing was uniform. Corduroy trousers, held up by a wide leather strap and buttoned at the knee, gave way to high-laced boots. A square-tailed coat of velveteen or jean over a black-spotted scarlet waistcoat, and a white felt hat. A brightly-coloured neckerchief was optional. For the most part illiterate, they nevertheless possessed great practical knowledge of the nature of rocks, soil and clay, and their skill in the various arts of tunnelling was considerable. With an almost unlimited capacity for continuous back-breaking toil they could, when necessary, endure sixteen hours of work with only short breaks for meals. They had a complete disregard for danger—in fact, the jobs in highest demand were those that were most perilous, presumably because of the exhilaration that such jobs entailed.

The navvies were described by the nineteenth-century reformer Edwin Chadwick as 'drawn together from all parts by thousands—most of them men of prodigious strength, violent passions, and ignorant to a fearful and almost incredible degree . . . having no home but the public-house by day and a barn or shed or temporary hut by night—having no other pastime after their hard work than drunkenness and fighting'.

The picturesqueness of their dress was surpassed only by that of their language, while their ferocious behaviour was equalled only by the vicious-ness of their drinking habits. One can imagine the plight of a remote village like Kilsby when several hundred of these men took up residence near by. The navvies showed respect and loyalty only to members of their own gang—their hands were turned against every other man's with the result that every man was against them. But there was little, if anything, that the men of a village could do about the orgies of drunkenness, fighting, riot and destruction that occurred when the navvies were in funds. When this was not the case, robbery, with or without violence, was often resorted to without the slightest attempt at concealment.

The average navvy's wage was 3s 0d a day—a princely sum for a labourer in those days. The tunnelling navvy, however, was a prince among princes who could earn no less than 4s 3d a day—an amount which, to an ordinary English working man, was beyond his dreams. Needless to say, such big money in circulation attracted plenty of people who were willing to help the navvy in spending it. When the work was to begin on a new line the racketeers were there before the navvies, building huts to be let and setting up shops. Constructed of mud and turf, the huts were made simply by running rafters from a hedge or bank to the ground and building up the sides with turf. The result was a lean-to room of about 12 feet square that could be let to two men at a rent of up to 6s a week. Often the owners of these huts were the contractors themselves. At a Committee of Enquiry into these conditions, the Rev. J. R. Thompson was asked

> 'Are they [the huts] at all equal to the class of houses which are generally occupied by the labouring population?'
> 'I never saw anything to be compared with them.'
> 'Will you describe them to the Committee?'
> 'I can only describe them as being built against this hedge or bank; the rafters sloping from this bank, and the sides and ends of turf. In some cases I believe they are just boarded inside, in some cases they are not; in some cases there is no partition; man, woman and child all sleep exposed to one another; sometimes a sort of temporary curtain may be thrown across,

which divides the sleeping apartment from what they call the kitchen, where the fireplace is.'

'Is that an accommodation with which they are satisfied?'

'In the case of families I know they have complained very much, and in one or two instances, where the individuals have been able, they have fitted them up a little better, and then put down stone and got matting, and so improved them that they seem tolerably dry, and in that case they are fair dwellings.'

The contractors also controlled the tommy-shops, from which all but the most provident of the men bought food, clothing, tobacco and supplies. The average navvy would not dream of spending his hard cash on anything other than hard liquor. For any other commodity he applied at the pay office for a 'ticket'. Tickets could be for any amount between 1*d* and 6*d* and they were used as cash—but were only acceptable at the tommy-shop. The amount each man had received in tickets was deducted from his fortnightly pay. The charges for goods in these shops were grossly exorbitant and the profits went into the contractors' pockets. The Rev. Thompson remarked that the men paid thirty per cent over the odds for the food bought at the tommy-shop.

'Do you know what they pay for meat?'

'6½*d* and 7*d*; and that is for the very coarsest joints.'

'What is the general price?'

'That is about the price for the best joints; . . . of course inferior joints are generally about 2*d* a pound less, and these are the joints principally bought up for the supply of the men.'

'They do not insist upon having the best meat?'

'They can only have what the shop has procured for them.'

'These labourers, if that shop were not there, having spent their money and wanted something during the week, if they went to the shop in Totnes and bought upon credit, they would be equally liable, I apprehend, to overcharge?'

'I do not think that any of the shops in Totnes would trust them. I think they would be silly to do so, speaking from my own experience. There is not an atom's worth of honesty among them generally. I have lent them money over and over again myself, and in not one instance have I been voluntarily repaid.'

In 1845, at the height of the railway mania, there were some 200,000 men building 3,000 miles of new railway in Britain, and, in the eighty years following 1822, 20,000 miles of line in Britain alone besides many thousands of miles more in the rest of Europe and America were constructed. It is not to be wondered at, then, that the engineers in charge of these vast projects tackled tunnels not so much singly as by the dozen.

6

Some Brunel Railway Tunnels

WHEN, IN 1835, Marc Brunel restarted work on the Thames tunnel he was obliged to look for another resident engineer to replace his son, for Isambard Kingdom was by then a fully-fledged railway engineer with responsibility for building the Great Western Railway between Bristol and London. Of the $118\frac{3}{4}$ miles of this line, the most difficult and trying section was that between Chippenham and Bath, for it was here that the engineer had to pierce the massive Box Hill with a tunnel nearly 2 miles long—by far the longest railway tunnel yet attempted. Built for double broad-gauge track the Box tunnel is 30 feet wide at the base of the arch, the crown of which is 25 feet above the rails. A total of 247,000 cubic yards of spoil were excavated during the four and a half years it took to build, and during this period one ton of gunpowder and one ton of candles were used every week.

In a world that still judged speed by the standard of the horse there was, not surprisingly, considerable opposition to the Box tunnel. The idea of propelling a train at high speed through a $1\frac{3}{4}$-mile tunnel on a 1 in 100 gradient was considered in some quarters as 'monstrous and extraordinary, most dangerous and impracticable'. The train would emerge with a load of corpses and even if the journey did not prove fatal 'no passenger could be induced to go twice'. A Doctor Dionysius Lardner, himself a competent authority on railway management, demonstrated on paper that if the brakes of a train failed as it entered the higher end of the tunnel it would emerge at the other end at the incredible speed of 120 m.p.h., at which rate, it was well known, no human being could breathe.

The Box and Woodhead tunnels

Work began in September 1836, and by 1840 a work force of 4,000 men and 300 horses was working around the clock to complete the tunnel on time. Eight working shafts were sunk; six were to become permanent ventilating shafts, of which the deepest was 300 feet. The shafts were excavated by horse-gins and an example of 'horse economics' as applied to gin driving has survived.

Cost of 67 horses per day

	£	s	d
Two quarters, 1 bushel beans @ 38s per quarter	4	0	9
Two quarters, 1 bushel oats @ 25s per quarter	2	13	1
Fifty trusses of hay @ £5 10s per ton of 40 trusses	6	17	6
Forty trusses of straw @ 8d per truss	1	6	8
Shoeing each horse, 1d per day per horse		5	7
Farriers expenses, 3d per day per horse		16	9
Stabling expenses, 3d per day per horse		16	9
Harness and repairs, 3d per day per horse		16	9
12 Stablemen @ 3s per day	1	16	0
18 gin-boys @ 1s 3d per day	1	2	6
18 gin-boys @ 1s 6d (night shift)	1	7	0
	21	19	4

or 6s 6$\frac{7}{10}$d per day per horse.

The eight shafts, which were each 30 feet in diameter and from 70 to 300 feet deep, were completed by the end of 1837, and the work of tunnelling was let out to two contractors. George Burge, a well-established railway contractor, took over all except the last half-mile at the eastern end. The latter was undertaken by two local builders, a Mr Brewer of Box and a Mr Lewis of Bath. This section was to be driven through the great oolite or Bath stone, and Brunel had this cut as a Gothic arch and left it unlined.

The immense quantity of stone, clay and earth taken from beneath Box Hill was, apart from the aid of gunpowder in breaking the rock, dug out of the facings by pick-axe, shovel and human sinew, and hauled to the surface by men and horses. The only other power used during the whole tremendous operation was the steam which drove the pumps to keep the works dry. The amount of water encountered must have reminded the engineer of his days under the Thames river-bed, especially during wet weather when water poured through the fissures in the ground. On one occasion, in 1837, the inrush was so great that the pumps could not cope with it and the water reached 56 feet up the shafts. The atmosphere in that gloomy candle-lit cavern was appalling. The choking stench of gunpowder fumes and dust mingled with damp air and the sweat and breath of men and beasts so crammed together in that restricted space that the swinging picks and shovels could be lethal to the unwary. Over one hundred lives were lost in that underground battle, fought continuously for nearly five years. The Box tunnel was completed in 1841 and opened to traffic in June of that year. It was described in a report to the Board of Trade as follows: 'The tunnel is 3,193 yards, rather more than 1¾ miles in length, in forming which eight large shafts . . . were excavated, of which Nos. 1 and 8 were afterwards enlarged into the openings or deep cuttings for the entrances of the tunnel at each end, so that six only remained when the work was finished.'

Not every passenger travelling between Paddington and Bristol could face the prospect of shooting through Box Hill like an arrow, amid sulphureous fumes and shooting sparks from an express locomotive; instead, in order to avoid these terrors, they would quit the train at Corsham or Box, post over the hill, and catch the next train bound for their destination.

Another of Brunel's tunnels on this line is worthy of note. During its construction, part of the ground slipped away, making it unnecessary to complete the top of one of the side walls. Brunel, therefore, gave instructions that it should be left as it was and planted with ivy so that it would resemble a ruined arch.

The 3-mile railway tunnel, on the Sheffield–Manchester line near the small Cheshire village of Woodhead, was built during the years 1838–45. The story of its building was hailed by a newspaper of the time as 'a wondrous triumph of art over nature' and 'the greatest engineering work of the kind which has yet been completed'. More recently, Terry Coleman described it in *The Railway Navvies* as a 'most degraded adventure' and 'a story of heroic savagery, magnificent profits and devout hypocrisy'. All these descriptions are accurate in their way for the Woodhead tunnel was a masterpiece of engineering achieved through much misery and loss of life.

The engineer to the undertaking was Joseph Locke, one-time pupil of, but by now rival to, George Stephenson, and from the very start of the work he experienced great difficulties. The tunnel was to penetrate the Pennines at a height of 1,000 feet above sea-level, and the first year was spent in surveying and preparing the site and making trial borings along the proposed line. The tunnel was to be driven from both ends and from five working shafts of an average depth of 500 feet. It was envisaged that the shafts would be sunk to grade within the first working year, but No. 2 shaft took four years to complete because of the enormous quantities of water that were met unexpectedly at that point. At one time 100,000 gallons of water *a day* were being pumped out of No. 2, in order merely to stop the level from rising. Such was the unpredictable nature of the rock that all but 1,000 yards of the tunnel's length of 3 miles 22 yards had to be lined with masonry. Working conditions were, if that was possible, worse than those below Box Hill. The flow of water from walls and roof was continual, the men working knee deep in mud and soaked to the skin. At several places the flow of water was so great that the pumps had to operate around the clock; it was estimated that during the six years of work 8,000,000 *tons* of water were removed from the workings along with 273,000 cubic yards of spoil. One hundred and fifty-seven tons of gunpowder were used to blast through the rock—an operation which, in such close, treacherous conditions, was the cause of many losses of life. Of these casualties, a considerable number were caused through the use of iron stemmers, or rams, to pack the powder into the drilled holes. The iron stemmers tended to spark against the rock when in use, thus causing the gunpowder to explode in its rocky barrel and drive the stemmer into the body of the unfortunate miner.

One example is described in the Enquiry report: 'William Jackson, miner, No. 5 shaft. He was looking over John Webb's shoulder, while he was stemming a hole charged with powder, when the blast went off, blowing the stemmer through Jackson's head and killed him on the spot.'

After a number of accidents with iron stemmers it was suggested to the contractors that copper stemmers, which were far less liable to cause a spark, should be substituted. But this idea was turned down owing to the high price of copper. When the Enquiry investigated working conditions at Woodhead, assistant-engineer Wellington Purdon stated that patent fuses were not used in blasting and, when asked by the committee whether the patent fuse was safer than the old type, he replied: 'Perhaps it is; but it is attended with such loss of time, and the difference is so very small, I would not recommend the loss of time for the sake of all the extra lives it would save.' Thirty-two men were killed, many more were seriously wounded and an unrecorded number died from tuberculosis, bronchitis and other diseases through the perpetual damp.

The list of non-fatal casualties treated by one of the two doctors employed by the company included: 21 compound fractures; 2 fractured skulls; 74 simple fractures; 140 severe cases of burns from blasting; contusions, lacerations and dislocations; one man lost both eyes and half his foot; 450 cases of trapped and broken fingers. Writing in 1846, Edwin Chadwick commented:

> Thirty-two killed out of such a body of labourers, and one hundred and forty (*sic*) wounded, besides the sick, nearly equal the proportionate casualties of a campaign or a severe battle. The losses in this one work may be

stated as more than three per cent of killed, and fourteen per cent wounded. The deaths (according to the official returns) in the four battles, Talavera, Salamanca, Vittoria and Waterloo, were only 2·11 per cent of privates; and in the last forty-one months of the Peninsula war the mortality of privates in battle was 4·2 per cent, of disease 11·9 per cent.

On 20 December 1845 the tunnel was declared open to traffic, and the workmen were rewarded with a dinner that included a roasted bullock of 'above ordinary size' and as much ale as they could drink.

The first Woodhead tunnel was, as we have seen, for single working, but the engineers, aware that a second line would be needed, had driven twenty-five side arches into the wall at intervals of 200 yards. No working shafts were required therefore when the second tunnel was begun in 1847. The nature of the ground now being known to the engineers, and with much improved access to the work-faces, the second Woodhead tunnel was constructed with far less loss of life through accident, although there were almost as many deaths through a cholera epidemic that struck the workings in 1849. The men, working in damp and dismal conditions and in close contact with one another in the tunnel, turf huts and crowded bar-rooms, were ideal victims of that then almost invariably fatal disease. For once the company shouldered its responsibilities by sending an ample supply of coffins to the works, with the result that the men became unnerved and abandoned the workings completely. When twenty-eight men had died, the epidemic abated and work on the tunnel was continued. Both the up- and the down-line tunnels remained in use for over one hundred years. They were both closed in 1954 when the third, double-line tunnel was built by British Rail.

The Battle of Mickleton Tunnel

The labourers who dug and hewed through ridge and hill, although treated by companies and contractors as cattle to be sacrificed on the altar of dividends, were capable of a remarkably pugnacious loyalty to their employers when the need for it arose. This was dramatically demonstrated in 1851 in an affair that became known as the 'Battle of Mickleton Tunnel'. The tunnel which was to carry the Oxford, Worcester and Wolverhampton Railway under the Cotswolds had been a source of difficulty and trouble since it was begun in 1846. In 1851, Marchant, the contractor, had a dispute with the company concerning the terms of his contract and the ownership of certain items of plant being used in the work. The company, considering Marchant to be in breach of his contract, dismissed him and gave the work to another contractor, Messrs Peto and

Betts. But when the latter arrived at the site with their work-force they found themselves confronted by Marchant with a strong army of his men. Skirmishes and a form of guerrilla warfare went on between the two sides for some weeks before the engineer to the Company—none other than Isambard Brunel—turned up in command of a large force of men. The ensuing battle is best described in the words of the *Illustrated London News*:

> For some time past an extraordinary degree of excitement has prevailed in a little village named Cambden, in Worcestershire, a place where the head offices of the contractors for the formation of a line of road through the Muckleton [*sic*] Tunnel are situated, in consequence of a party warfare being carried on between the officers of the Oxford, Worcester, and Wolverhampton Company and a contractor of the name of Marchant, with whom the company had some differences on the subject of the completion of his contract; in consequence of which it was resolved that that part of the line should be also taken in hand by Messrs Peto and Betts, the contractors for the whole line, with the above exception, between Oxford and Worcester. To this end, upon several occasions during the past month, Mr. Marchant has been requested to desist from keeping on any workmen; but, he having refused, physical force was resorted to. The agents of Messrs Peto and Betts were ordered to collect about 500 men, and march them on last Sunday night to Muckleton tunnel, in order to be the first on the ground, and prevent Mr. Marchant's men from pursuing their work on the Monday. On reaching the Worcester end of the tunnel, Mr. Cowdery, with 200 men from Evesham and Wyre, with their pickaxes and shovels, was met by Mr. Marchant, who dared any one of Messrs Peto and Betts's men to pass the bridge, on pain of being shot, Mr. Marchant himself being well supplied with pistols. Mr. Brunel, engineer of the line, finding expostulation unavailing, gave peremptory orders for Messrs Peto and Betts's men to proceed and take everything on the line. A rush was then made by the men, which for a few seconds was repelled with great force by Marchant and his men, and the consequence was that several heads were broken, and three men had their shoulders dislocated. A man in the employ of Marchant having drawn pistols, he was seized upon and his head nearly severed in two. Marchant then left Messrs Peto and Betts's men for an hour in undisputed possession of the ground; but at the expiration of that time he returned with some three dozen policemen from the Gloucester constabulary, and some privates of the Gloucester Artillery, accompanied by the two magistrates of the place, who immediately commenced reading the Riot Act. At this juncture a *mêlée* had taken place on a high embankment, and here several broken limbs were the result of the conflict. About four o'clock, Mr. Charles Watson, of Warwick, arrived with upwards of 200 men, and the Great Western Company also sent a similar number, in order to expel Marchant. The magistrates here told Marchant's men to commence their work, but no sooner was the order given than Messrs Peto and Betts's agents were directed to stop work, even by force, if necessary.

Eventually, Brunel reinforced his army by transferring squads of navvies from other points on the line and even from the works of the Birmingham and Oxford at Warwick. With 2,000 men under his command, he advanced in an encircling movement that surrounded Marchant and his men; finding themselves outflanked, they surrendered and the engineer was acknowledged as victor of the Battle of Mickleton Tunnel. The report goes on to relate that while Marchant and Brunel were coming to an amicable settlement 'a small batch of navvies again met, and one of them had his little finger bitten off, and another his head severely wounded'.

7

Tunnels of the French
and the Americans

THE MOST FAMOUS tunnels in the world are those that pierce the most formidable physical barrier in the continent of Europe—the Alps. Romans, Goths, Huns, Carthaginians, French, Germans and Austrians have marched through the natural openings of this high mountain range and military engineers have built fine roads through them, but it has never been possible to keep those roads open all the year round. The great wave of trade that started in the eighteenth century was, by the middle of the nineteenth, still breaking fruitlessly on that massive granite barrier, while the two halves of Europe could only communicate overland by means of packhorses and mules. Many engineers were consulted to devise a means of passing a railway through the Alps, but all pronounced it impractical. No locomotive could haul a train up the gradients that would be necessary to reach the passes—and even if such an engine could be contrived it could not run when there was snow. The only alternative was to keep the level of the railway at a sufficiently low level to avoid the snow, and this meant tunnelling. The Box Hill tunnel in England had been driven from five shafts and both ends simultaneously and its 2-mile length had taken five years to excavate. A tunnel driven under a mountain could not, by its very nature, be driven from shafts and to drive through the hard Alpine rock from the two ends alone would, it was estimated, take at least ten years per mile. Then again, how was a tunnel without shafts to be ventilated, especially a tunnel built to accommodate express locomotives? The difficulties seemed so enormous as to be scarcely worth consideration. However, they *were* considered in the 1850s when the problem of linking the French and Italian railways was again examined. At this time, travellers from France to Italy had to detrain at Mondane and, with their luggage, make a 50-mile journey over the mountains.

The Mont Cenis tunnel

To eliminate this irritating and uncomfortable journey, the French engineers decided to drive a double-track railway tunnel through Mont Cenis for a distance of 8 miles—a decision that brought forth this comment from a contemporary journalist: 'Dismiss the notion! Crossing a mountain that way is beneath the dignity of humanity. We declare for truth and daylight! If we must visit our neighbours let us do so openly, not crawl in among them like a burglar from a cellar grating, or a rat popping up out of a dry sewer.' No rock-drilling machinery was then available so it was intended to drill all the bore holes by the same method that the Romans had used, that is, one man hammering a 'jumper' that was held and turned by another. This method is slow and laborious to say the very least; the hardest operation of all being 'overhand stopping' for work on the roof—one man holding the 'jumper' to the roof while two others struck at it upwards with sledge-hammers. Work started in 1857, using the Austrian system, holes for gunpowder being drilled to depths varying between 18 and 36 inches. Progress amounted to an average of 9 inches a day—a rate of advance which, if continued, would have kept the engineers busy for about seventy-five years.

The compressed-air drill

But in 1849, Jonathan Couch of Philadelphia had designed a mechanical drill, worked by compressed air. It activated a steel drill which delivered a series of hammer-strokes to the rock, while at the same time twisting it around to clear the hole of débris. Many mechanical drills were developed from this idea but no practical method had been devised for compressing air. Appropriately enough, it was Germain Sommeiller, engineer in charge of the Mont Cenis project, who solved the problem. Working in conjunction with the engineers Grandis and Grattoni, he produced a machine, called a '*corp de bellier*', which utilised the water-power available at both the tunnel portals. Not only was it capable of powering the drills, but the exhaust air from it ventilated the heading, thus solving the problem of dust which would, surely, in time, have brought the work to a standstill.

When the machinery was installed in 1861 the rate of advance rose dramatically and continued to do so year by year. By 1870 the combined yearly advance of the two headings was 5,364 feet, or nearly ten times the rate of the advance in the first year. Later, Sommeiller designed a 12-ton railway carriage on which was mounted up to nine drilling machines, each of which could be adjusted for direction by means of screws. In the absence of vertical shafts, all the débris had to be hauled to the tunnel entrance, and by 1865 no less than eighty horses were employed on this work alone.

In the pilot tunnel, three shifts of sixty men worked around the clock. Thirty of these operated the drilling-rig, while the remainder did the blasting and mucking. Eighty feet back from the work face another 230 men excavated the tunnel to its full size, while, farther back still, eighty more men and ten boys built the lining.

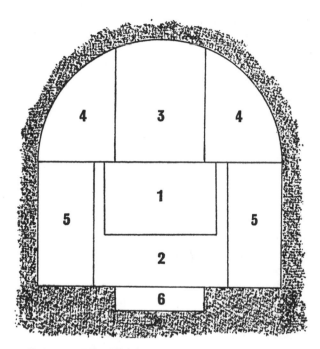

Figure 15. Mont Cenis tunnel—order of excavation.

The tunnelling was carried on in a cycle of five operations (*see* Fig. 15). First, the drill carriage was hauled up to the work-face and 80 holes were drilled in a series of concentric circles. This operation took twelve hours. Next, the blasting party advanced and filled the holes with cartridges of black blasting-powder. The holes were then plugged with wood through which fuses were threaded. Next, the ventilating plant was set to work; and, finally, after the charges had been exploded, the muckers took over to clear away the débris. The advance heading of about 7 feet by 7 feet was then opened out to run along the bottom of the full bore. The atmosphere inside the tunnel was choking from the start, becoming worse as the work-face penetrated deeper into the mountain. When the miners had reached a length of 3 miles, conditions grew intolerable. The ventilating system kept the actual work-face clear by driving the smoke and dust farther back, but there it just hung in the air mingling with the breath of hundreds of working men and horses and the fumes from gas- and oil-lamps. One English visitor to the works complained that, although in the tunnel for less than an hour, he emerged coughing and spitting, 'as black as though I had dined on lampblack'.

On Christmas Day 1870, after over thirteen years' continual work, the two sides met and the tunnel was opened to traffic three months later. So accurate was the original survey that the difference in level between the two headings was 12 inches, and that in alignment only 18 inches. The total cost of the tunnel was £3,000,000 and twenty-eight lives. The Mont Cenis tunnel is 7·9 miles long and, at its maximum depth, nearly a mile beneath the earth. The tunnel made it possible to travel from Paris direct to Turin in under eighteen hours.

An unnecessary tunnel

Early in the nineteenth century, the rage for canal-tunnel building reached the United States. Not to be outdone by the Europeans, the directors of the Schuylkill Navigational canal in Pennsylvania insisted on tunnelling through a ridge of shale at Orwidsburg Landing, near Auburn. There can be little doubt that the main purpose of tunnelling at this point was to promote public interest in the canal, for a tunnel could have been avoided by locating the route of the canal farther along the ridge. In addition, the ridge was only 40 feet higher than the canal level and a cutting would have been more easily achieved and far simpler to maintain and navigate than a 450-foot tunnel.

This was demonstrated in 1855 when the tunnel was completely removed to transform America's oldest tunnel into an open cut. Three years after this almost unique engineering work—an unnecessary tunnel—a second tunnel was built in the United States. This was at the summit of an old canal that ran for a distance of 72 miles along Swatara Creek to Pine Grove in the Schuylkill coalfields. Its length was 720 feet and its cost was $30,464. Driven through slate it was brick arched at each end to a distance of 150 feet.

The 'Great Bore'—the Hoosac

Another landmark in the history of American tunnelling was the Hoosac tunnel, which opened in 1876 after twenty-one years' work. A tunnel was first mooted in about 1820, when it was proposed to dig a canal from

Boston to the Hudson river, to provide a more direct route from eastern Massachusetts to the Hudson, western New York and the Great Lakes. The route of the proposed canal was, however, blocked by the imposing Hoosac mountain—an obstruction that the engineers calmly proposed burrowing through! Surveys for the scheme were carried out in 1825, but, probably because of the immense difficulties the task presented, the scheme was shelved. By 1848 the rapid expansion of the railways had rendered the Boston–Hudson canal scheme obsolete, and plans were made to cover the route with a railway instead of a canal. Although the plan had changed, the Hoosac mountain had not, and it now towered athwart the route between Greenfield and Williamstown where the new line was intended. At the end of 1848, a Charter was granted to the Troy and Greenfield Railway Company to construct a tunnel under the Hoosac. The first surveys were carried out in 1850.

The tunnel was to be for two tracks, 4¾ miles long, 24 feet wide and 22 feet high. It was to be driven from the two ends and from a shaft near the middle. The cost was estimated at $1,948,557, and it was reckoned that the work would take 1,550 days.

Work started in 1851 with the aid of a complicated drilling machine, weighing some 70 tons. It comprised a number of drills, mounted to drill a circle of holes around a core that subsequently could be blasted out. The steam-powered machine actually penetrated the mountain to a depth of 10 feet before work was suspended, due to the Company's inability to raise capital. It is difficult to see how the engineers planned to supply the drills with steam had the machine penetrated much farther. (Indeed, short of following up behind the works with a large portable boiler plant, in itself an impossible operation in the circumstances, there was no way of doing this.) When work was resumed more conventional methods were resorted to.

In 1854 the Company secured a State Legislature Loan of $2,000,000 and awarded the contract, for $3,500,000, to E. W. Serrel & Co. in the following year. For reasons that are not altogether clear, Serrel & Co. relinquished their contract in 1856 and the work was again put out to tender. This time the fee was raised to $3,880,000, and the new contractor advanced the tunnel from both ends to a total length of 4,250 feet and sank the central shaft to 306 feet. In 1861 the contractors were refused a certificate of payment and they, too, retired from the scene. Nothing more was done until the State foreclosed its mortgage on the incomplete tunnel and took over the work itself. The following six years showed slow progress at a cost that was out of all proportion to results. Call after call was made on the State Treasury as the cost per yard of tunnel increased. By this time the 'Great Bore', as the Hoosac tunnel had become known, was a State scandal. Rumour had it that money destined for the work was being diverted to the bank accounts of state officials and as these rumours gathered strength they were taken up by the newspapers.

The State Commission took over the workings in 1862, but no advance was made on the tunnel for nearly three years. This time was spent in damming the Deerfield river to provide water power to work the air-compressors and in completing the shaft which was oval-shape, measuring 27 feet by 15 feet. In 1865 tunnelling was recommenced on three faces, each measuring 15 feet wide by 6 feet high. By the end of the year, 634 feet had been tunnelled at a cost of 9,522 man-days and 11,195 pounds of gun-powder. It is recorded that 153,436 drills had been dulled to make this gain. In 1867 the extensive timber installations in and around the base of the working shaft caught fire. For the thirteen men working at the bottom of the shaft there was only one way to the surface and that, acting as a flue for the flames, was a column of fire. The men were either suffocated or burned to death.

In 1866 Thomas Doane, the state's chief engineer to the Great Bore, was approached by a New York agent, T. P. Schaffner, who intimated that he might be able to speed up progress through the Hoosac. Schaffner brought with him some samples of a new explosive which, demonstrated by him on a mass of rock outside the tunnel, produced a most shattering effect. The name of this new blasting agent was nitro-glycerine or, as it was then known, nitroleum—an amber-like fluid, discovered by Sobrero in 1847. Alfred Nobel first attempted its application as an explosive agent in 1864; but until he invented the process of stabilising it by mixing it with siliceous earth, in 1868, its use, especially in confined spaces, was a most risky operation. As we have seen, however, human life rated third to money and time in those days of engineering, so Doane, once having witnessed the destructive power of the new explosive, had no hesitation in employing it to speed up the Great Bore. Charges were set off by electric fuses and, as may be expected, accidents were numerous. On one occasion a workman walking through the tunnel in rubber boots that insulated him from the ground, became a human battery. At the work-face he picked up the bare end of a wire attached to a charge with the result that he was blown to pieces.

Although the use of the new explosive did improve the rate of advance to some degree, progress remained poor—while the rumours of bribery and corruption increased. At the end of 1868 more than $7,000,000 had been spent and only a third of the course had been covered. The state decided to put the remaining 3 miles of tunnel to contract and the new contractor, W. and F. Shanley, of Montreal, agreed to finish the work for the sum of $4,594,268. The new contract was signed early in 1869 and for the first time since the work was begun, eighteen years earlier, it was pushed ahead with energy and enthusiasm.

The average monthly rate of advance increased immediately from 47 feet to 126 feet; indeed, during one month in 1873 an advance of 162 feet was made. The final holing-through blast was fired on 27 November 1873. Enlarging and lining took another year, and in February 1875 the first train passed through the Hoosac tunnel. It consisted of a box car and three flat trucks bearing all the officials concerned with the promotion and financing of the project, and the engineers responsible for its construction. The total cost of the Hoosac tunnel was never accurately arrived at. There were a number of official estimates and they varied from between $10,000,000 and $20,000,000. The middle figure of $15,000,000 is the same as the total cost of the Mont Cenis tunnel and that was almost twice as long. Despite the muddle and inefficiency that made the driving of the Hoosac an epic in inertia, the tunnel made history in many ways. It was the first project in the United States that employed compressed-air drilling and the first time that nitro-glycerine was used in tunnelling.

New techniques and demand

During the twenty-two years it had taken to pierce the Hoosac mountain in America, tunnelling techniques all over the world had advanced greatly, and mechanical rock drills had undergone considerable development and improvement. As a consequence, tunnels were built for new purposes.

A pneumatic despatch railway was constructed beneath the streets of London in 1865. Built to carry letters and parcels between Euston Station and the General Post Office in the City, the iron-lined, brick tunnel was 4 foot 6 inches wide at floor level and 4 foot high. Steam-driven suction fans created a vacuum in the tunnel by centrifugal force, while the cars were fitted with rubber flanges forming an airtight seal to the tunnel's iron lining. At the stations, air was excluded from the tube with hermetically sealed spring doors. No drivers were required, and speeds were controlled by varying the amount of air in front of the car through a series of air conduits. The cars travelled at an average speed of 35 m.p.h., with a top speed of 60 m.p.h. on the incline of Farringdon Street. Great difficulties were encountered, however, in keeping the tunnel airtight and in maintaining the seal between car and tube, and the tunnel was abandoned and bricked up in about 1875. It still exists and is now used as a Post Office cable duct.

The problems of Chicago's water

Up until the year 1861, the rapidly growing city of Chicago obtained all its water from Lake Michigan; unfortunately the lake was also used as a receptacle for much of the city's waste, with the result that the water around the shores became badly polluted. It was therefore decided by the water authorities to build a 5-foot-square tunnel, to reach 3,522 yards out into the lake to retrieve pure water. A timber 'crib' was first built on shore. Forty feet high, the crib was pentagonal in shape, each of the sides being 58 feet long; in the centre a 'well', 22 feet in diameter, ran through it. When completed the crib was towed out and sunk into position; this was accomplished by filling the annular space between the well and the outer pentagon with pieces of rock. With the crib in position on the bed of the lake, a cast-iron cylinder, 9 feet in diameter, was lowered into the well until it, too, was resting on the bottom of the lake. Water was pumped out of the cylinder and, when it was dry, men went to work inside to excavate the clay. The result of this was that the cylinder sunk under its own weight to form the working shaft. When it had reached a depth of 23 feet below the bottom of the lake, a tunnel was driven out from it towards the shore at an average rate of $9\frac{1}{2}$ feet per day. Meanwhile tunnelling was proceeding from the shore side, also from a cast-iron shaft. The clay was found to be stiff enough to stand unsupported for about thirty-six hours, allowing ample time to put in the permanent lining without preliminary timbering. The most serious trouble met by both headings was the unusually large quantities of methane, which caused much singeing and burning when it ignited. The miners soon learned to detect the vicinity of these pockets, for they produced a hollow sound when the clay was struck with a pick. When a pocket was found a hole was bored into it. The gas was then lit and the men stood well back until it burned itself out.

When the two headings were 100 yards apart, work was stopped on the main headings while a pilot was driven to connect them in order to check level and alignment. An error of a little over 7 inches was found and this was easily corrected over the 100 yards that still remained to be dug out. The tunnel took just three years to complete and cost $457,815. The timber crib was enclosed with granite, and on its top a fine lighthouse was built.

Inspired by Chicago's solution to its water problems, the city of Cleveland, in 1869, followed suit by starting the construction of a 2,200-yard-long tunnel under Lake Erie for the same purpose. Work began with the sinking of the shore shaft which, at 63 feet, ran into a large pocket of methane gas. The foreman and one of his miners went down to see what could be done; *what* they did will never be known for certain for the result was a violent explosion that blew the two men off the face of the earth. (It is said that the foreman, an experienced coal-miner, struck a match!) A blast of hot air and flame rushed up the shaft, straight into the faces of the rest of the shift who were looking down to see what was going on; most of them were severely burned.

Plate 9. The Wapping end of the Thames tunnel, as it was when originally opened for pedestrians.

Plate 10. Wapping station, with a steam train emerging from the tunnel. Both this illustration and the one above appeared in the *Illustrated London News* of 8 January 1870.

Plate 11. The rear view of a Greathead shield, first used to drive the Tower subway.

Plate 12. The Tower subway, the first iron-lined tunnel and the first tube railway, opened in 1870.

Plate 13. The Niagara tunnel—advancing a top heading and bench simultaneously. The railway at the upper level was laid on planks supported by hangers suspended from timbered supports. Illustration taken from *Appleton's Cyclopaedia of Applied Mechanics*, 1892.

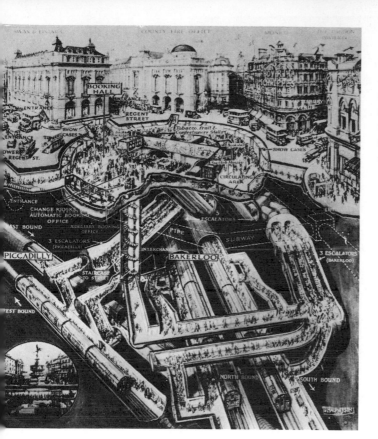

Plate 14. Piccadilly Underground
station. The huge booking hall
and accompanying maze of
tunnels were excavated through
a single shaft only 18 feet in
diameter.

Plate 15. 'Legging' through the Blisworth tunnel, on the Grand Union canal (*c.* 1900).

Plate 16. Workers operating medium-weight rock drills.

Plate 17. Sectional model of the Orange–Fish tunnel. The black markers indicate the location of the seven working-shafts, the deepest of which has been sunk 1,245 feet into the mountain.

When, eventually, the shaft reached grade, or full depth, the 5-foot-square tunnel was started, but from the very first the heading had to struggle through water, mud, quicksand and gas. When the heading had advanced about 200 yards all these enemies poured into the tunnel at the same time, making further progress impossible. The heading was sealed off. It was then decided to work around this impossible ground by turning off at an angle of 20° for a distance of 40 feet and then to turn back and pick up the original course. This was done and the slow, difficult advance continued. Then, on the night of 29 April 1871, the working shift, crammed shoulder to shoulder in the confined working area, suddenly found themselves ankle-deep in water that was rising rapidly. This was alarming, for the water came not from the work-face but from behind them: the men threw down their tools and made for the shore portal at full speed. Two hundred yards back from the work-face they passed the source of the trouble. Water was spurting through the lining in various places with considerable force. So great was the pressure behind it that nothing could be done to stop the leaks. The heading was abandoned and sealed off.

While the shore party was suffering these tribulations another heading was advancing to meet them from a shaft in a wooden crib out in the lake. Conditions there were no better. At first soft clay was met, but that soon gave way to almost liquid mud. A tunnelling-shield was obtained but the ground being so soft, the shield sank under its own weight, making it impossible to maintain the correct level. Finally it became jammed on some rock and was removed from the heading. By then, however, firmer ground had been reached and it was possible to advance by the fore-poling method (*see* page 14). Progress was slow but steady until the heading had reached to within 20 feet of the face of the abandoned shore tunnel. Here disaster struck again. The whole of the work-face suddenly ballooned out and a mass of clay, water and gas swept the entire gang off their feet and surged into the heading. With their lights put out and with timbering falling down on them, it is a miracle that the eighteen men who made up the shift were able to escape from that low, narrow heading. Bending to avoid the roof and guiding themselves along the wall, they reached the shaft just in time and all of them climbed up the ladder to safety.

Nothing could be done in either of the flooded headings until powerful pumps were obtained and installed at both shaft-heads. Then the headings were cleared and work restarted to excavate that last 20 feet. A great deal of gas remained hovering in the arch of the lake heading, and this was dealt with by a simple, if rather dangerous, method. Wearing lighted candles on their helmets, the men crawled along the heading igniting the gas which exploded harmlessly above their heads. When contact was made with the shore heading it was found to be reasonably dry but the lining had settled into the mud—in places up to 5 feet deep. Nothing could be done to rectify this, so two bulkheads were built to seal off the flooded section, which was over 800 feet long. The tunnel was then re-routed, two new headings being driven off and turned to run towards each other, parallel to the original line. The ground encountered was much firmer and the tunnel was completed without further difficulties.

The Chicago disaster

Many miles of water tunnels have been constructed under the Great Lakes since those of Cleveland and Chicago, but the ease with which the latter was built has never been repeated. A terrible disaster occurred in 1909 during the building of another lake tunnel to supply water to Chicago. The wooden inlet crib, some 7,500 feet out in the lake, caught fire while one hundred men were working on it. Before a boat could approach the blazing timber, sixty men had died. Some, trapped within the crib, were burned to death; others drowned after being forced into the water, while many who ran along the heading to escape the flames were later found suffocated.

The tunnellers' three main enemies, fire, water and lack of air, had combined to make this one of the worst disasters in tunnelling history.

8

New Developments
and the
St Gotthard Tunnel

THE TRAGEDY OF the building of the Thames tunnel, its enormous cost in terms of life and money, and its lack of commercial success deterred both promoters and engineers from attempting to tunnel under water in the years that followed. Marc Brunel's tunnelling shield was a brilliant invention but it proved to be far from foolproof. It was not until 1869 that another sub-aqueous tunnel was attempted and this again was under the Thames.

Peter Barlow and his shield

The engineer to that project was Peter William Barlow who can claim to be the father of modern tunnelling. Barlow's shield, improved by J. H. Greathead, whose name it now bears, permitted a tremendous advance in tunnelling techniques, enabling tunnels to be driven at greatly increased speeds through a wide variety of ground with both safety and economy. Born in 1809, Barlow specialised in bridge-building and it was during the construction of a suspension bridge across the Thames at Lambeth that the idea of his tunnelling-shield came to him. During the process of sinking or forcing vertical cast-iron caissons into the clay of the river, it occurred to him that iron cylinders could be driven horizontally and, in suitable soils, could be employed for tunnelling under a river-bed. In 1864 Barlow patented his cylindrical shield, and in 1868 his proposal to use it to build a foot-tunnel under the Thames at Tower Hill was accepted. Work started on the 1,350-foot tunnel in 1869 and was finished in under six months—a time that would be considered remarkable even today. Barlow's shield was an open-ended cylinder, about 8 feet in diameter and nearly 3 feet in length. The cutting-edge was a cast-iron rim, braced with six wrought-iron spokes extending half-way towards the centre where they joined an inner hexagon. The cylinder was slightly tapered from the cutting-edge towards the rear, in order to reduce skin friction with the soil through which it was to pass. In the rear of the cutting-rim and placed on the angles of the hexagon were six screw-jacks, each $2\frac{1}{2}$ inches in diameter and abutting against the completed lining of the tunnel, by means of which the shield was propelled through the ground. Instead of lining the tunnel with brickwork, Barlow instituted another important advance in tunnelling, the cast-iron lining, prefabricated in segments. The lining was of slightly smaller outside diameter than the inner diameter of the shield so that it could be assembled and bolted together within the tail-end of the shield itself.

A 60-foot shaft was sunk on the Tower side and the shield lowered into it. From there it was pushed through the clay at an average rate of 9 feet per day. The three miners employed in the shield first excavated the clay through the central hexagonal opening. When sufficient space had been cleared, one man entered the cavity and enlarged it to allow a second man to join him. With the opening enlarged still further, the third man got to work and the clay was removed to the full diameter of the shield. The six screw-jacks were then brought into operation to move the cylinder forward for up to 2 feet. In its rear was erected one section of lining, 18 inches long, and made up of three segments. At the end of each forward shove, the tail portion of the shield still overlapped the forward end of the completed tunnel, thus providing a closed end within which the next lining-ring could be safely erected.

Except when advancing through very soft ground, Greathead's shield leaves a ring-like, or annular, space between the outside of the tunnel lining and the burrow made by the shield—a space that can in time cause settlement of the ground above and resultant damage to buildings. Greathead overcame this disadvantage by the invention of his grouting pan. By this method, liquid cement, or grout, is forced under air-pressure through holes in the lining which are subsequently closed with screw-plugs.

The Tower subway cost a mere £16,000 and was opened to traffic in 1870. Passengers were conveyed through it in a cable-hauled omnibus, powered by steam-engines at both ends. Although, like Beach's first 'tube' railway, it was never a commercial success it was remarkable for the simplicity, speed and economy of its construction. It was the first tunnel to be driven by modern methods for, in essence, modern tunnelling-shields are the same as the one invented by Barlow and improved by Greathead. The

Tower tunnel continued in use until the opening of Tower bridge, in 1894. The tunnel was then closed to the public and used to carry water-mains under the river.

Alfred Beach

While Barlow's shield was making its easy way under the bed of the Thames, Alfred Beach, an American inventor and engineer, was working, quite independently, on very similar lines in New York. Beach's tunnel ran under Broadway from the south-west corner of Warren Street to a point near the south side of Murray Street. It was not constructed because a tunnel was particularly needed at this point, but rather to demonstrate the practicability of two ideas of Beach's concerning tunnelling under city streets: first, that pneumatic propulsion was the only practical method in an unventilated tunnel; secondly, that a tunnelling-shield could be used in constructing subways, thus avoiding the interference with traffic that would be caused by the cut-and-cover method.

Beach's shield was very similar to that of Barlow's and it may be that he copied the latter in many respects. There can be no doubt, however, that he was the first to apply hydraulic jacks to push the shield forward. Work started in 1869 and the tunnel was completed by the following year. It was 8 feet in diameter and its crown was 21 fee tbelow ground. The passenger car was blown forward by air pressure one way, and 'sucked' back by a vacuum on its return. The system was opened to the public in 1871 at a charge of 25c a ride; it became one of the biggest tourist attractions in the city. Yet in spite of this it was a commercial failure and was closed and bricked up two years later.

The St Gotthard tunnel

The second piercing of the Alps was started in 1872 to take the Zürich–Milan railway underneath the St Gotthard pass. To drive a tunnel 9¼ miles long through solid rock would, a few years previously, have been considered an impossibility, but with the new drilling machines and explosives available in 1872 and, above all, with the successful tunnelling of Mont Cenis as an example of what could be done, the project seemed quite feasible; in fact, when the Swiss Central Railway Company invited tenders seven firms competed for the contract. This was to prove un-fortunate, for the keenness of the competition resulted in the winning contractor, Louis Favre of Genoa, taking on the job for £1,898,945, a figure that allowed for no contingencies whatsoever. Favre undertook to complete the tunnel in eight years and was required to put up £320,000 which he would forfeit if the work was not finished one year after the agreed time. If the tunnel was completed *before* 1 October 1880 he was to receive a bonus of £200 for every day saved, but, if it was not finished by that date, he was to pay the Company £200 a day for the first six months and £400 a day thereafter for the next six months. If the tunnel was then incomplete still, Favre was to forfeit his £320,000 deposit.

The significance of Favre's contract

In agreeing to these harsh terms Louis Favre signed not only his own death warrant but that of hundreds of his men. The driving of the St Gotthard tunnel is a supreme example of what a man will suffer for money. For seven years Favre, continually threatened by the deadline that was overtaking him, worked his men without thought for their health or their lives. Pressured to the point of desperation by the railway company, he would spend days on end in the choking atmosphere of the tunnel, driving the men until they literally dropped in their tracks. Eventually, through the sheer exhaustion that was the result of an incessant struggle with men, money, materials, machines and the mountain, he was to pay the penalty.

The conditions under which the men toiled was far worse than any encountered before, or since, and the casualty rate was frightful; accidental explosions, falls of rock, and mishaps with machinery accounted for an average of a death every two weeks, in addition to hundreds of serious injuries. Hundreds more died from disease—silicosis, bronchitis, pneumonia and 'miner's anaemia'—and from the attentions of the parasitic worm ankylostome. Three or four months of continuous employment in these conditions usually brought sickness to a man, while twelve months work could make him a chronic invalid—that is if he survived.

An American journalist visiting the works in 1875 later wrote an account of his visit for Harper's *New Monthly Magazine*:

> As we rushed by dripping walls, and saw here and there ghoul-like figures with dim lamps hiding behind rocks or in deep niches, I involuntarily recalled what our conductor had said of a glimpse of the bowels of hell.
>
> It was impossible to speak and be heard. I might as well have addressed myself to the granite walls or the tunnel as to have attempted a word to either of my companions.
>
> The air was so thick that lights could not be seen twenty yards ahead of us, and we all walked close together for fear of being lost or tumbling into some subterranean hole.
>
> Far ahead of us we heard the dynamite explosions, sounding like heavy

mortars in the midst of battle. In some places where we were walking the water was nearly a foot deep, and again it came through crevasses about our heads like April showers . . . [The men's] food is extremely limited in quantity, and is wretched in quality, consisting largely of polenta, or a sort of Indian meal porridge. Meat they never taste at all. They are contented to receive their forty or fifty cents a day for hard work, if they can only escape wounds and death from the bad gases and the thousand accidents to which they are liable every moment of their lives in the tunnel. Alas! they do not escape, for every week records its disaster, either from explosions and flying rocks, falling timbers and masonry, or railway accidents, breaking machinery, etc.

The tunnel was driven from two headings and was started in September 1872. At both the north and south portals huge, complicated installations had to be built—housing for over 4,000 men, hospitals, shops, stores, workshops, stables, foundaries and smithies. The cost of these alone amounted to £240,000, over one-eighth of Favre's total budget! Then the rivers had to be diverted and harnessed to provide power for four turbines of 250 h.p. each that worked the air compressors for the drills. This preliminary work took two years and while it was being carried out tunnelling was started with hand tools at a rate of about 2 feet 6 inches a day.

Favre had decided on a modified version of the Belgium method to drive the St Gotthard tunnel (*see* Fig. 16). The top heading, which varied

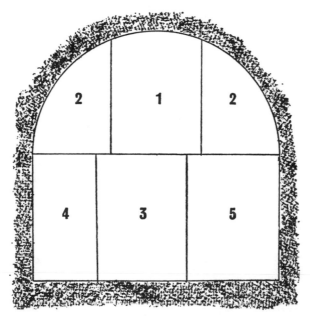

Figure 16. Saint Gotthard tunnel—order of excavation.

from 53 to 75 feet square, was excavated by a drill carriage of six drills that made up to twenty-five holes to a depth of 4 feet. These were charged with dynamite and fired. This took two and a half hours, while the clearing of resulting débris took three and a half hours. By working round the clock an average advance of 13 feet a day was maintained. Behind the work-face

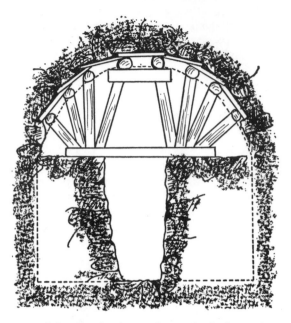

Figure 17. Saint Gotthard tunnel—method of strutting roof.

and strung out for nearly 1,000 yards, men with drills broke the heading out to full size and built the masonry lining. The work was wet from the very start. In both headings water issued from the walls in thick jets, sometimes with enough velocity to knock a man down. Water gushed out of the drill-holes with such force that the charges had to be wedged into place before they could be blasted. At one point in the tunnel, water was flowing in at a rate of 3,000 gallons per minute. As the blood-soaked paths of the two passages penetrated farther and farther into the rock, the temperatures in the headings rose to 106° F, with the result that men were dropping by the dozen and having to be carried or dragged back into the open air. By 1879 conditions had become so hideous that working time was reduced to five and a half hours a shift. By this time the contract price had been exceeded and Louis Favre had spent some £300,000 of his own money to keep the work going. On 15 July 1879, Favre, broken physically as well as financially, was himself struck down and was buried in Göschenen cemetery among hundreds of workmen who had died before him.

The St Gotthard tunnel was completed by Favre's two assistants and the two pilot headings met at 11.00 A.M. on 29 February 1880. The deadline date of 1 October 1800 came and went and over a year later there were still 3,319 men working in the tunnel, breaking it out and lining it with masonry. By the time the work was finished Favre's firm had exceeded the

Figure 18. Saint Gotthard tunnel—position of wagon tracks.

contract price by no less than £590,000 and in addition had forfeited the deposit of £320,000.

The St Gotthard tunnel was the ruin not only of the Favre firm but also of the Favre family. To show their gratitude to the man who had sacrificed his fortune and his life to the tunnel, the Swiss Central Railway Company awarded his only daughter an annuity of £8 a month.

The total cost in money of the St Gotthard tunnel amounted to £2,300,000, or £143 per linear yard. The cost in lives was 310, with 877 seriously invalided. The tunnel remains a major work of engineering, though its successful completion relied entirely on the tenacity of the engineers and the powers of endurance, albeit enforced, of the workmen. Unfortunately it cannot be said that these men endured such dangers and sufferings for the sake of high rewards. Life must indeed have been cheap in the Alps in those days for men to do such work, in such conditions, for 'fifty cents a day' in order to keep body and soul together with 'polenta . . . a sort of Indian meal porridge'.

9

The Severn Tunnel

WHEN BY THE 1870s the British railway system was virtually complete, engineers turned their attention to closing the gaps between the trunk lines, most of which were the result of natural barriers. At this time the River Severn divided the Great Western Railway into two sections connected by steam-ferries that were subject to tides, storms and fogs. In 1871 it was decided to take the railway under the river and in the following year parliamentary approval for the scheme was obtained. The Severn tunnel was the first submarine tunnel of any great length to be built; it consists of $4\frac{1}{2}$ miles of double-track tunnel, $2\frac{1}{4}$ miles of which run beneath the river-bed.

Work began in 1873 with the construction of a number of substantial and comfortable cottages for the skilled workers, a hospital, nurses' home and a school. Mission halls and coffee taverns were built for the workforce, and a road to the nearest village to give access to shops. In the March of the same year a start was made on sinking a shaft, 15 feet in diameter and 200 feet deep, at Sidbrook on the Welsh side. At the very outset of this work the engineers received a taste of the troubles that were to beset them over the next fourteen years. When the 'old shaft', as it became known, reached a depth of 45 feet, water poured into the shaft at the rate of 12,000 gallons an hour. Work came to a standstill while pumping machinery was installed to deal with the inundation. When this was done, work started anew, but a few feet lower down another spring burst in on the works to deliver a further 27,000 gallons an hour. To deal with this a 9-foot stroke-plunger pump was installed, powered by a 40-inch Cornish beam engine. When the shaft reached grade, a heading, 7 feet square and known as 'seven feet heading', was driven under the estuary on a rising gradient so that any water would flow back into the shaft to be dealt with by the pumps. So much water was met with that it became necessary to sink another pumping shaft some 27 yards from Old Shaft. Lined with iron, this became known as Iron Shaft. Iron Shaft was connected to the Seven Feet heading by a cross-cutting fitted with a door.

Calamities and dangers

For six years the work continued smoothly, until by 1879 only 130 yards separated the two headings. Then the real trouble began. The pilot tunnel, being driven towards the land from the shore shaft on the Monmouthshire side, tapped a great spring, and water poured into the tunnel at the rate of 360,000 gallons an hour. It flooded through Seven Feet heading in a great wave; poured into Old Shaft, overcoming the pumps in seconds; and rushed through the heading towards the miners who were working some 2 miles away. Alerted by the roar of the water, the men escaped through the door in Iron Shaft in the nick of time. There were then five pumps on the Monmouthshire side and three on the Gloucestershire side, but they were far from capable of clearing the immense amount of water that now poured into the workings.

In the face of this calamity, the eminent engineer Sir John Hawkshaw was put in charge, and the experienced tunneller Thomas Walker was given the contract to finish the work. New and more powerful pumping machinery was put on order, and three new shafts were sunk in readiness for it. One of the pumps was 38 inches in diameter, with a 9-foot stroke, and was the largest in the world. As soon as the new machinery was installed, pumping began at a rate of 460,000 gallons an hour and the water level dropped from 150 feet to 30 feet. Here it stuck. It was then discovered that an iron door in the heading, 1,000 yards from Old Shaft, had been left open by the men as they escaped from the flooding and it was estimated that a further 150,000 gallons of water were flowing from this source to join the 300,000 gallons from the Great Spring. It was decided to close this door, and a diver named Lambert, together with two assistants, was sent along the tunnel to do this. The journey along the heading was a dangerous one. Lambert stationed one of his men at the bottom of the shaft to pay out his line, while the second was put half-way along the heading to pull the pipe forward and feed it to the diver. There were no underwater lights in those days and Lambert had to feel his way through pitch darkness, almost redigging the 1,000 yards of tunnel, so cluttered was it with timber, rock and overturned trucks. With only 30 yards between him and the door, Lambert was checked by his air-line, which had become twisted around the various obstacles along the way. There was nothing for it but to give up the attempt and return. The journey back was even more difficult and dangerous. Lambert had no way to instruct his assistant to take the line in again, so he had to coil it up himself as he went along.

At his second attempt he wore a new type of diving dress that had recently been invented by a naval officer, named H. A. Fleuss; this apparatus carried its own supply of oxygen and did not, therefore, require an air-pipe. This time Lambert reached the door, but was unable to close it because of the skip-rails that passed through its opening. On November 10 he made his third attempt, this time managing to pull up the rails with a crowbar before shutting the door. The pumps were now able to master the water and by mid-December the tunnel was dry. A week later the heading through which the Great Spring had entered was explored and the flow was walled in with timber. Work was next started on clearing up the mess in the heading, preparatory to restarting tunnelling, but almost immediately a severe snowstorm began—cutting off the works from the rest of the country. Here indeed was a crisis, for no coal could be brought up for the pump boilers. When the supplies on the site were exhausted the stocks of timbering were burnt and, at the end of two weeks' isolation, even furniture was being fed into the furnaces.

Before Walker took charge, the men worked nominal eight-hour shifts. The first started at six in the morning, firing a round of shot in the face of the heading, mucking out and then finishing. By this system three shifts a day were worked, half an hour being allowed for a meal while the men were still in the tunnel. As all the blasting was done with dynamite—the fumes of which are dangerous—the men were allowed to wait for the smoke to clear after each blast before returning to work. To save this wasted time, Walker introduced a ten-hour system of two shifts by which the men came out of the tunnel for their meals immediately after firing the charges. The air was then clear by the time they went down again. The men started work at six o'clock (A.M. and P.M.) and worked for three hours, coming out of the tunnel for breakfast from nine to ten. Going down again they worked until one o'clock, when they came up for dinner. Then from two until six they worked to complete the ten-hour shift. In 1881 there was a strike among the men for a return to the eight-hour shift, but this collapsed after two days when Walker made it perfectly clear to the men that he would close the workings indefinitely rather than meet their demands.

The breakthrough of the Severn

In April the Severn itself broke in through the tunnel roof, flooding an extensive portion of the works—the breakthrough occurred at a point where the river was only 3 feet deep. To find the hole, Walker resorted to a method that he could hardly get away with today. Without telling his men what he was about, the contractor ordered a number of them to link hands and wade into the river. Eventually Walker observed 'one of the men suddenly popping down out of sight, and being pulled out [of the water] by those who had hold of his hands on either side'. Walker had found the hole. The leak was then closed with a load of clay.

On 26 September 1881, the two headings met exactly—there being no errors either in level or alignment—and work continued in breaking out and lining the heading, a task that took another two years.

On the afternoon of 2 December 1882, a workman rushed into Walker's office 'with a face exhibiting the most complete signs of terror' and shouting at the top of his voice, 'The river's in, the river's in!' Walker hurried to the top of Old Shaft where, in his own words, he found

between 300 and 400 men, evidently in the greatest terror and distress. Some had lost part of their clothing; hardly one of them could speak from exhaustion; and they were anxiously watching for the arrival of the large cage, that was bringing up a further batch of men.

Every man was panting for breath, and excited to the last degree with fear. I must say that my heart sank, and I feared the worst; at that moment the cage arrived at the top with ten or twelve men, and a foreman . . . I turned to him eagerly and said:

'Lester, what did you see?'

'I see nothin', sir.'

'What is it, then?'

'I don't know, only the river's in.'

'Where were you working?'

'In No. 8.'

'And you saw nothing?'

'No, it was beyond me.'

I turned to another, and said:

'And what did you see?'

'I see nothin', but the river's in.'

'I think you are a pack of fools,' I said.

'I think a five-pound note will cure it all.'

Just then the foreman of the Cornish pumps came up, and said, 'I cannot understand it; there is no more water at the pumps.'

I walked over with him . . . to the edge of the river, and found the water from the discharge culvert the same colour as usual; so I walked back to the pit, and said, 'I am going below: who is going with me?'

J. H. Simpson and Jim Richards jumped into the cage with me, and the signal to lower was given. On arriving at the bottom we found it *perfectly dry*, and four or five men sitting quietly on a piece of timber scraping their boots.

It was expected by both engineers and promoters that the tunnel would be open to traffic by 1885 but these hopes were thwarted by the Great Spring which, although imprisoned, was awaiting its opportunity to break out.

In October 1883 the vicinity of that prison was reached and the most elaborate precautions were taken to by-pass it; but all in vain, for on October 10 the water burst into the workings at a rate that made the inundation of 1879 look like a dribble. This time the torrent roared into the tunnel at 27,000 gallons per *minute*, or 1,620,000 gallons an hour, against a pumping capacity of 660,000 gallons an hour. With foresight born of his previous experience, Walker had built in flood doors at intervals along the finished tunnel, and once more the indomitable Lambert went down to close them. This done the pumps began to master the rising level of water in the shafts.

Walker, by now, must have felt that whatever further troubles Fate had in store for him, they would not come in the form of water. He was wrong. On the night of October 17 a furious storm broke over the Bristol Channel, bringing in its wake a gigantic tidal wave which towered over the sea-walls, swept through them, and fell upon the low-lying country-side beyond. The water swept into the works, extinguished the boiler fires and poured down the shafts into the tunnel. Eighty-three men were at work below ground at the time, and their only means of escape was cut off by the cascade falling down the shafts. With the level of the water rapidly rising they climbed up the timber staging until they found an overhead cavity high in the rock. A small boat was lowered down the shaft and, by sawing a way through the timbers, its crew managed to reach the men and rescue them.

After the two weeks' pumping it was possible to restart work and efforts were directed to settle accounts with the Great Spring once and for all. A separate heading was driven below the spring, its water tapped into it and pumped up. To avoid having to pump the Great Spring for ever it was again walled up in the hope that the water would find itself another channel but within three weeks it was exerting a pressure on the tunnel lining of 54 pounds per square inch. Before long, bricks were shooting out of the lining like cannon shells, to be followed by powerful jets of water that threatened to flood the tunnel again. There was nothing for it but to install permanent pumps, and this was done by the middle of 1886. Since then the pumps have worked continuously and they have never brought up less than 23,000,000 gallons a day. On one occasion they were recorded as having raised 34,061,378 gallons, weighing over 152,000 tons, in twenty-four hours—an inflow rate that would fill an average size public swimming-bath (say 70,000 gallons) in a matter of three minutes.

On 1 September 1886 the first goods train passed through the Severn tunnel, to be followed three months later by the first regular passenger train. The tunnel took nearly fourteen years to construct and at one period there were 3,100 men employed at wages varying from 10*d* an hour for foremen to 2*d* an hour for boys. Together with 36,794 tons of Portland cement, 76,400,000 bricks were used.

In 1929 a section of the Severn tunnel was subjected to 'cementation', a process that consists of boring a number of holes at intervals in the brick lining and injecting cement under pressure through the holes to fill any voids on the outside of the walls and to form a jacket around the lining. As a result the Severn tunnel is one of the driest in the country, although the amount of water negotiated during its building (in addition to its running under a tidal estuary) suggests that it should be the wettest. With its regular use by passenger traffic, the travelling time between Bristol and Cardiff was reduced by one hour—and the initials G.W.R. ceased to stand for Great Way Round.

Sub-aqueous
Tunnelling

I N SPITE OF the advances that had taken place in civil engineering during the nineteenth century, the Hudson river between New York and Jersey City could, until 1908, only be crossed by ferry. A bridge across the mile-wide, 70-foot-deep river was even then beyond the capabilities of engineering and the nearest one was 145 miles upstream. In 1873 a railway engineer named De Witt Clinton Haskin promoted the Hudson Tunnel Railroad Company, raised $10,000,000 and started to burrow under the soft mud of the river by means of compressed air. Although compressed air had not been used before for tunnelling, it had been used for caisson-sinking as early as 1830 by the British engineer Thomas (later Lord) Cochrane. Cochrane had invented and patented the first air-lock to give access to the airtight caisson comprising a closed-end cylinder with a door at each end; the door into the compressed-air chamber opened outwards and an outer door opened inward.

The principle

The principle of driving through mud either horizontally or vertically can be simply described. If the air pressure inside the tunnel or shaft is raised and maintained to that of the water trying to get in, the flow of that water is stopped. The pressure inside must match that of the outside exactly. If the pressure is too low the water will still come in; if it is too high any weakness in the ground may give way and force a hole through to the river-bed. If this happens the internal compression is lost and the chamber will be instantly flooded.

The compressed air method had been used by the American engineer James Buchanan Eads to sink the caissons of his bridge over the Mississippi, and the progress of this work was followed closely by Haskin. At first he considered using compressed air behind a cylindrical tunnelling-shield, but then he had the daring idea of merely maintaining an air pressure in the heading. This, together with ordinary timbering, would, he reckoned, be sufficient to hold back a shell of iron plates which, in turn, would hold back the mud until the permanent masonry was built. Haskin's first step was to design and build his air-lock. This took the form of an iron boiler, 15 feet long and 6 feet in diameter, with a door at either end.

The beginning of the Hudson river project

In 1874 work started on a vertical shaft on the Jersey City side of the river at the foot of 15th Street, but it was hardly begun before it was stopped again by the first of a series of injunctions. The power behind most of these orders was considerable, for they were obtained by the Delaware, Lackawanna and Western Railroad Company which, as operators of the ferry near the route of the proposed tunnel, had much to lose by its success. Although the railway company was in the end defeated, it managed to hold up work for five years.

By the end of 1879 the shaft had reached its full depth of 60 feet, and Haskin's air-lock was lowered into it and fixed into a concrete bulkhead. The compressing machinery was installed and connected, and work started in a pressure of 17–20 pounds per square inch. Round-the-clock shifts were worked, each of the three gangs comprising forty men. Sure enough, the pressure held back the mud and water long enough to allow the timbering to be erected; this was then replaced by a series of iron rings which, in turn, were bricked in with the permanent lining.

When passing from a compressed-air chamber to the normal atmosphere, a man must be 'decompressed' gradually, otherwise he may soon feel the unpleasant symptoms of caisson disease or, as the men call it, 'the bends'. The symptoms are caused by the tissues and blood becoming saturated with nitrogen through the pressure of the air. When pressure is reduced the nitrogen effervesces and forms bubbles which slow down the circulation of the blood. In severe cases the heart will stop beating. To avoid this the air-lock is decompressed slowly, depending on the pressure of the air. Sometimes men may have to stay three hours in the air-lock. In the 1870s caisson sickness was rare and little was known about it. Consequently the men in the Hudson tunnel suffered greatly from it although, as the pressures used were not very high, there were no fatalities.

The tunnelling was done in three stages. A top heading, about 6 fee high was first driven and lined with iron plates braced with timber. As this heading advanced, a centre heading, also about 6 feet high, was dug, but lagging behind the top heading. This was lined in the same way, while behind it the final 6 feet were excavated. As the heading advanced, therefore, it resembled three large steps. The excavated spoil was removed by mixing it with water in a large trough and forcing the resultant liquid to the surface through 6-inch pipes by the air-pressure in the heading. Work went smoothly and progress was good until July 21, by which time a distance of 300 yards under the river had been completed. Then, on the morning of that day, a leak was discovered in the gallery. There was nothing particularly serious or unusual in this. One of the men approached the leak with a sandbag, to stop it when the hole suddenly widened; the compressed air in the shaft fled with a hiss, and a great jet of water and slime poured into the workings. The roofing plates, no longer supported by air pressure, collapsed, and within seconds the entire tunnel was half filled with oozing mud. What happened next is a story of the coolest heroism.

Woodland and his miners

There were twenty-seven men in the heading at the time of the blow-out, including the shift foreman Peter Woodland. The men, led by Woodland, managed to scramble back to the air-lock. Seven men got into it while Woodland stood outside the entrance marshalling the men through the door. At this point a roof-fall half closed the air-lock door and at the same time thoroughly blocked it so that it was impossible to enter the chamber.

Inside the lock the men panicked. The farther door, opening inwards, could not, of course, be opened against the air pressure in the heading which, although reduced through the 'blow', was still being fed by the compressors and was considerable. In the centre of the farther door was set a small inspection window of very thick glass and in the chaos of the moment Woodland could see what had to be done. Although he was fully aware that his orders would cost him his life, he roared at the men in the lock to be silent, then ordered them to smash through the window with a piece of iron. This would release *all* the air in the heading which would then immediately fill up, but it would enable the seven men in the lock to open the door and climb the shaft. Even the panic-stricken seven hesitated to take such an action but Woodland, as calm as if he were directing operations at the tunnel face, ordered the glass to be broken. This was done and as the seven fortunate ones made their escape Woodland and twenty workers died in the darkness.

Several months were spent in blocking the leak, pumping out the heading and putting the lock to rights. Haskin now realised that it would not do to rely completely on air pressure to hold up the roof, and a new method of work was adopted. This was to drive a small pilot tunnel about 10 feet in advance of the main tunnel. The pilot tunnel was lined with stout steel-plate pipe, and the plates forming the preliminary shell of the main tunnel were braced against it. On 20 August 1882 another blow occurred to flood the heading, but this time no lives were lost. At that point the Company ran out of money and in view of the last accident Haskin could obtain no further backing.

This was but one of a series of long stoppages—it was not until 1908 that the first train ran under the Hudson river. Altogether the construction spread over a period of thirty-five years, and during this time the techniques of tunnelling developed considerably. One of these developments was the invention of the Brandt rock drill, in 1876. Driven by water instead of air, it did much to increase the speed of tunnelling. The drilling stem was hollow, with a bit furnished with four teeth so arranged that a hole was driven slightly larger than the stem. While the earlier, air-driven drills hammered at the rock, the Brandt drill bored the hole and was capable of drilling to a depth of 39 inches in fifteen minutes. Rotating at a rate of ten revolutions a minute it was powered by two small cylinders driven by water under high pressure. Exhaust water was expelled through a pipe around the centre of the drill which was thus kept cool, while the water jet also washed out dust and débris from the bore. The machine was forced forward by a hydraulic ram abutted against a beam wedged across the heading with a force of over 10 tons.

The Brandt drill was first used on a long tunnel in 1880 when work began on the 6½-mile-long Arlberg tunnel that gave Vienna a railway route to Paris via Switzerland. The line climbs a series of steep gradients before entering the tunnel, the first 2⅓ miles of which are on a gradient rising 1 in 500 before falling at 1 in 66 for the remaining 4 miles. Some water was encountered but the two headings met exactly three years after work was begun—an average advance of 9 yards a day compared with the 2½ yards a day accomplished by the Mont Cenis tunnellers.

We have seen how, in 1870, James Henry Greathead improved and developed Barlow's tunnelling-shield and used it with great success in driving the Tower subway. Even though this tunnel had been dug in a remarkably short time, Greathead had experienced some difficulty when wet ground caused a collapse of the working-face in front of the shield. It was therefore with great interest that he followed Haskin's attempt to tunnel under the Hudson river with compressed air alone, and it soon became obvious to him that the Haskin method and the Greathead method needed each other for complete success.

The Greathead shield

Haskin's troubles lay in the difficulty of supporting the tunnel walls, a difficulty that could be overcome with a shield; Greathead's problem was to hold back the soft work-face, and this could be done with compressed air. On this basis he designed the Greathead shield—the prototype of all subsequent tunnelling-shields. The chance to test this invention came in 1886 when Greathead was engaged to build what was to become the nucleus of London's great underground railway system, a tunnel running between the Monument in the City and Stockwell, south of the river. At first called the 'City of London and Southwark Subway', it later became the 'City and South London Railway'. This, the first 'tube' railway, comprised twin circular tubes, 5 feet apart and from 10 feet 2 inches to 10 feet 6 inches in diameter. The shield was 11 foot 4 inches in diameter and 6 feet long. In the upper part of the rear bulkhead was set an airtight door leading into an air-lock built into the cylinder. Needless to say, the engineers still had to deal with the major problem of compressed-air tunnelling, that is, to maintain the correct air pressure that would hold back the soft ground without blowing through to the surface.

The ground was excavated 2 feet at a time, the shield being advanced by six hydraulic jacks. The lining of cast-iron rings, each 19 inches long, was erected in segments bolted together, while the annular space between lining and tunnel was grouted under pressure. At various times eighteen shields were employed to drive the City and South London on eight different working-faces. The 3½ miles of underground railway were advanced in the remarkably short time of four years, during the entire course of which there was not one fatal accident. The completion of the work marked the beginning of two eras—that of modern sub-aqueous tunnelling and that of the underground railway.

As steam locomotion is impractical on an underground railway, it was first intended to draw the trains through with cables hauled by stationary engines placed at intervals along the tunnel. This scheme was abandoned, however, in favour of electric traction. Fourteen electric locomotives, each of 25 h.p., were used to haul five-carriage trains at a speed of 25 miles per hour. By the end of 1890 a regular three-minute service was in operation. One of the first passengers on the 'tube' recorded his impressions thus:

> The up and down lines are carried in separate tunnels placed at such a depth under the surface of the roads as to avoid interference with one another, or with the sewers and other underground structures. The comparatively small but ample platform accommodates the waiting passengers who have at present only the bare white walls and arched roof to gaze upon.

By-and-by a rumbling is heard; it becomes a roar, and then swells into a rush as the advancing train, emitting electric sparks apparently from the region of the rails, emerges from the black-mouthed tunnel. Each train is composed of three long cars, one of them for smokers. There is no distinction of class, and all alike are comfortable. The atmosphere of the subterranean station is no doubt somewhat close, but not unpleasantly so.

While Greathead was making his successful way under the River Thames, a Swedish engineer, Captain Lindmark, was driving through waterlogged ground by the simple method of freezing it first! Siberian miners had for hundreds of years taken advantage of low temperatures to tunnel through wet ground,[1] and artificial freezing had previously been applied to the sinking of shafts. In 1884-6 Lindmark drove a 758-foot-long pedestrian tunnel beneath the ridge that divides Stockholm into two parts. The tunnel, 12 feet 8 inches high by 13 feet 2 inches wide, passes through coarse gravel containing water and has very little cohesion. The gravel was frozen solid by a dry-air machine delivering 25,000 cubic feet an hour. The material was then excavated as though it were rock.

The demolition of Flood Rock

A truly gigantic feat of sub-aqueous tunnelling was carried out in 1885 for the purpose of demolishing Flood Rock, a 9-acre mass of granite near Long Island Sound, New York. A shaft was sunk on the shore to a depth of 64 feet below sea-level, and from this 4 miles of galleries were driven out into the rock in all directions. The roof of the tunnel, which varied in thickness from 10 to 24 feet, was supported by 467 pillars, each being 15 feet square; to accomplish this some 80,000 cubic yards of rock were excavated, leaving 275,000 cubic yards to be blown up. Over 13,000 holes were drilled in the roof and pillars averaging 9 feet in depth and 3 inches in diameter; these were charged with 110 tons of explosive and the whole lot instantaneously fired by electricity. The resulting explosion lifted the sea in the area to a height of over 100 feet and Flood Rock was no more. The cost of this spectacular tunnelling operation cannot be ascertained, but it must have been far less than building a lighthouse and maintaining it in perpetuity.

The Glasgow District tube

The world's second 'tube' railway was built by the city of Glasgow in 1891-7. This was a 6½-mile loop-line of twin tubes known as the 'Glasgow

---

[1] This may account for the claim by the Russians to have been the first to use chemical freezing when they constructed the Moscow underground system in 1935.

District Subway'. The soil through which it passes is of a wet sandy nature and, under the River Clyde, comprises almost liquid mud. Four Greathead shields were used to drive the two bores in opposite directions from St Enoch Square. In spite of the loose nature of the soil, the work under the land made good progress and presented no serious difficulties. When the Clyde was reached, however, great troubles were encountered.

During the first 80 feet of under-river advance no fewer than ten blow-outs occurred, the worst of which, on 24 February 1894, sent the entire timbering of the heading up into the river leaving a hole in the river-bed 24 feet square and 16 feet deep. After each blow-out the resultant holes were plugged with clay bags and the tunnel pumped dry, but after pumping the tunnel for the tenth time the contractor gave up the struggle and resigned. His successor, George Talbot, tackled the job with painstaking care. To avoid further blow-outs he had the river-bed regularly inspected by divers, and if soil was washed away by the tide he had the spot re-inforced with clay bags. Air pressure within the tunnel was regulated to match the depth of the water in accordance with the tides. As the tide came in the pressure was increased and vice versa. It had been observed that all the blow-outs occurred during meal-breaks, nights or weekends when the tunnel was vacant, and it was correctly deduced that pauses in the work were dangerous. Talbot, therefore, ordered round-the-clock shifts, seven days a week and with no meal-breaks.

His decision to work his men on Sundays created much opposition in strictly Presbyterian Glasgow, but the fact remained that no more blow-outs occurred and the work was speeded up considerably. The first 80 feet of sub-aqueous tunnelling had taken five months to build but Talbot completed the second, parallel tube of 410 feet in three-and-a-half months. The application of compressed air to tunnelling brings with it an added danger, in addition to blow-outs. Fire in compressed air burns more fiercely and spreads more rapidly than it does in a normal atmosphere— the greater the pressure, the greater the danger. Thus it was that a small fire that broke out in one of the unfinished tubes under the Clyde swept through the piles of timbering and was soon out of control. Fifteen men were trapped at the heading and, with the fire cutting off their air supply, were soon nearly suffocating in the dense smoke. Fortunately Greathead's grouting-pan was still connected to a supply of compressed air, and this kept the men alive until they were rescued. Talbot was aware that it would take days to put out the fire, so while operations to achieve this were under way, a desperate attempt was made to reach the imprisoned men. With a gang of his most experienced tunnellers, Talbot made his way along the parallel tube to the point opposite the estimated position of the men. Here the rescuers battered and tore through the iron lining and then began to tunnel through the 5 feet of soft soil that separated them from the burning heading. Checks were made on the height of the tide, and the air pressure was adjusted to hold up the walls and roof of the makeshift passage. At last they reached the tube and with sledge-hammers smashed an opening through the cast iron, knowing full well that when they did so the variation of air pressure might bring about the unlined tunnel's collapse. At last, after twenty hours of continual work, they broke through and the fifteen distressed men were brought out.

Even when Talbot had driven well beyond the river he was still having trouble with water. First, the tunnel ran into the bottom of a disused, waterlogged quarry that flooded it for nearly a mile. Then, a few weeks later, it passed over a stream that forced its way up through the floor of one heading which had to be bricked up and temporarily abandoned. Later a heading, using compressed air, was driven towards it from the opposite direction. When the headings met, pumps were installed to keep the tunnel dry.

After nearly six years' work, most of which was on a round-the-clock basis, the Glasgow District tube was completed and went into operation on 21 January 1897. The single cars were cable-hauled for many years, being converted to electric working in 1935. At first a flat-rate fare of 1*d* a journey was charged, irrespective of distance, with the result that many people spent their spare days in travelling back and forth for the sheer novelty of the ride.

The completion of the Hudson tunnel

In 1889 yet another attempt was made to complete the tunnel under the Hudson river started by Haskin in 1874. This time British capital was behind the scheme, and two renowned British engineers were put in charge of the work: Sir Benjamin Baker and John Fowler, who had constructed what is still considered as one of the engineering wonders of the world—the Forth bridge. With the arrival of these two giants, poor Haskin was demoted to the position of superintendent of the works. A Greathead tunnelling-shield was used, with cast-iron segments for lining instead of Haskin's brick and sheet steel. Good progress was made up to the middle of 1891 when, with 1,500 feet separating the two headings, the work again ground to a halt through lack of money. In spite of the fact that only another $650,000 was needed to complete the $4,000,000 project, the extra capital could not be found and the works were again closed up. The situation remained thus until 1899 when the entire works were sold to a new company. Again the tunnel was opened and pumped dry, only to be closed up once more through lack of funds. Then, in 1902, another company was formed to finish the work. This time it was well supplied with money, and the Hudson river tunnel started moving again under the direction of Charles M. Jacobs, a tunneller of many years' experience. After a little uneventful progress Jacobs ran into very difficult ground comprising a mixture of soft, wet earth and hard rock, the latter, unfortunately, forming the bottom of the heading while the loose ground formed the roof. Blasting the rock at the bottom of the tunnel involved disturbing the wet ground above with the almost certain danger of blowing it out or bringing it down. Jacobs's answer to the problem was as ingenious as it was daring.

Five huge kerosene-fired blow-pipes were brought up to the work-face and played directly on to the wet soil for a period of eight hours, the shield meanwhile being hosed to prevent it from warping. The result was that the silt was transformed into firm clay which, supported by air pressure of 38 pounds per square inch, was able to withstand the shocks of blasting. After firing and mucking the rock, the dried-out clay was excavated, the shield moved forward, and the whole process repeated.

The northward-bound tube was completed in March 1904 while work on the south tube continued until September 1905; at one point a blow-out occurred which created a fountain of silt that shot over 40 feet above the surface of the river. On another occasion a workman named Richard Creegan, while attempting to stop a blow-hole with a bale of hay, was himself blown into the hole where he was stuck in the mud. Then, as his mates ran to rescue him, he was shot like a cannonball, through the silt of the river-bed, through 15 feet of water and up to the surface of the river. He was rescued and brought ashore quite unharmed. The first train ran under the Hudson river in February 1908—thirty years after the work was first begun.

Tunnelling at the End
of the
Nineteenth Century

IN COMMON WITH most other branches of civil engineering, that of tunnelling had its Golden Age in the nineteenth century during which all the major developments of techniques were made, a vast number of miles was burrowed through earth and rock, and some of the world's greatest tunnels driven. It is therefore fitting that the end of the century saw the start of the most notable and difficult tunnelling work ever undertaken. This is the Simplon tunnel that runs for $12\frac{1}{3}$ miles under the Alps, passing below the Simplon pass at a depth of over 7,000 feet.

The Simplon tunnel

It has been said about many trying tunnelling projects, and about the Simplon in particular, that had engineers known what they were in for they would never have started in the first place; but, on the other hand, as Rolt Hammond reminds us in *Tunnel Engineering*, 'all experience shows that if there is one thing certain about tunnelling it is its uncertainty. In many cases everything appears to be going well, and then there will be a sudden change of ground or a sudden and disastrous flow of water may hold up operations for a long time.' In the case of the Simplon tunnel these difficulties were only minor ones compared with some others that were experienced. Water was indeed encountered in large quantities, but of such a high temperature to make it impossible to work without bringing huge quantities of icy water in from the mountain streams to cool the working-face. At one point this water was gushing into the workings at a rate of 10,500 gallons a minute and at a pressure of 600 pounds per square inch; at another, the rock pressure was so great that the tunnel arching had to be constructed of masonry 5 feet 6 inches thick.

Agreement to build the tunnel was reached between the Swiss and Italian governments in 1895, and in August 1898 work was started under the direction of Alfred Brandt, who, as we have seen, was the inventor of the Brandt rock-tunnelling drill. Instead of a single shaft containing double tracks, Brandt decided on two separate parallel single-track tunnels, 55·8 feet apart at their centres. This was a stroke of genius for, knowing through experience gained from the building of the St Gotthard that ventilation would be a serious problem, he proposed driving the two headings simultaneously, connecting them at 200-metre intervals by cross-headings. Large electric fans installed at the portal of one gallery pushed air through one or more of the cross-headings, which were fitted with doors to control the flow, and back down the other. As it happened, the difficulties met with in the construction of the Simplon tunnel—especially those concerned with ventilation—would have been insurmountable had Brandt decided to work by the old, conventional, single-heading method. Three Brandt drills were used at a time against the rock-face, each drill boring four holes in the space of two hours. While the holes were being cleared and charged with 'gelatin'[1] the drills were taken back to a safe place. The charges were then detonated and the resultant débris loaded on to narrow-gauge trains. The drills were then brought up for the next advance. Over 4,000,000 charges, weighing 1,350 tons, together with over 3,000 miles of fuse were used, and at one time 2,000 men were employed on the work. The complete cycle of drilling, charging, detonating and mucking took about five hours and the average advance was 18 feet a day.

The first troubles were experienced on the Italian side when the heading had advanced some 2¾ miles into the mountain. Here water was encountered in double the quantities that had put a stop to work in the St Gotthard tunnel in 1873. Brandt had anticipated water and had cut a trench throughout the length of one of his headings to channel the water down the gradient to its mouth, but he had not expected such huge quantities. Water poured out of the working-face, overflowed the trench, and transformed both headings into fast-flowing underground streams some 3 feet deep. Work was stopped while engineers racked their brains in vain for a means to stop the torrent when suddenly it subsided of its own accord. Although the workings were never quite free of water it was possible to continue the advance.

[1] An explosive substance more powerful than dynamite, made by dissolving collodion-cotton in nine times its weight of nitro-glycerine.

Another, even more serious hazard, came from the enormous weight of rock pressing down on the roof. Cave-ins presented a constant threat, and when they occurred timbering was often crushed to splinters and steel girders had to be used to support the roof and walls. Made of rolled steel, 16 inches thick with 6-inch flanges, they were bolted to massive timber beams 20 inches square, but even they could not withstand the pressure. Eventually the spaces between these frames were filled with concrete and the advance went on.

Fiery heat and boiling water

On the Swiss side the work was well ahead of schedule and the Swiss engineers tunnelled their way well beyond the agreed limit and down the Italian decline. They, too, ran into water but with the difference that their supply was very hot. The temperature in the heading rose rapidly to 115° F. The engineers, believing the heat to be local, urged their men on to greater efforts in order to pass the hot region as soon as possible, but, although the rate of advance increased, the atmosphere grew even hotter as the heading penetrated deeper into the mountain. When 5 miles had been traversed, the heat in that confined space was like a baker's oven. The fragments of blasted rock were too hot to pick up with the bare hands. After a couple of hours in the gallery even the most robust of men fainted away and a regular shuttle system was operated to ferry unconscious tunnellers to the open air.

When the temperature reached 125° F the engineers, convinced that it could not possibly increase further, urged the men on to cooler conditions as a hungry army might be encouraged by the promise of food. But, incredibly, the heat continued to increase in step with the slow painful advance until, in 1904, the incoming flow of water was at a temperature of 138° F, that is, near boiling-point—while the rock walls of the tunnel were at 131° F. Work could not continue under such conditions and the heading was abandoned and sealed off with a pair of iron doors. Work continued on the Italian side until 24 February 1905, when the Italians broke through into the Swiss portion of the tunnel, releasing the hot water that had accumulated there which then escaped down the decline on the Italian side. On 25 January 1906 the first train passed through, and by the summer of the same year the tunnel was completed. The error in direction between the two headings was found to be a mere 8 inches, while that in level was 3½ inches—a degree of accuracy that can only be described as astonishing in a tunnel over 12 miles long.

The Simplon tunnel is still the longest continuous railway tunnel in the world, and is likely to remain so for some time to come. As a feat of engineering it is monumental and yet it was carried out with surprisingly little loss of life considering the hazards and difficulties that were encountered. This was due to the careful attention given to the health and welfare of the workmen. Both contractors and engineers had learned from the frightful death-roll in the St Gotthard, and the most stringently enforced precautions were taken to protect the health of men subjected to sudden changes from extreme heat to extreme cold. Dressing-rooms were built into the tunnel where men could change into dry clothes on leaving the works, their wet clothes being dried by steam-heat before they were needed again. The intestinal worm that caused so many deaths at St Gotthard was defeated in the Simplon by means of adequate ventilation: for every single cubic foot of air supplied to the St Gotthard, 25 cubic feet were supplied to the Simplon.

The problem of fresh air

While the Simplon tunnel was under construction a great deal of attention was being given to the problem of ventilating under-mountain railway tunnels. Because of the thickness of the roofs of these tunnels it was not practicable to supply them with ventilating shafts and with adverse wind, conditions inside them could be very bad. A case in point occurred in the single-line tunnel through the Apennines at Pracchia (*see* Chapter 12). A great deal of the traffic going through these tunnels was hauled by heavy steam-locomotives. The Pracchia tunnel, 9,000 feet long and with a gradient of 1 in 40, was subject to wind blowing in at the lower end and if, at the same time, a heavy train ascended the gradient, conditions became very serious. The engine, working full-steam to overcome the steep gradient, discharged a great quantity of dense smoke in addition to steam, and this was liable to be carried along with the train. Sometimes, heavy trains with two engines would climb the gradient very slowly with slipping wheels due to the wet rails, and conditions in the tunnel would become uncomfortable to say the least. A heavy passenger train, with a royal party on board, once arrived at the upper portal with both drivers and both firemen unconscious on the footplates; on another occasion a passenger train emerged with nearly all its passengers insensible. Before a satisfactory ventilation system was installed in this tunnel, both drivers and firemen had to wrap yards of woollen cloth around their heads before entering the lower portal.

The Loetschberg tunnel

Shortly after the opening of the Simplon tunnel, work was started on another major passage beneath the Alps. This was the Loetschberg tunnel, of just over 9 miles, only 430 yards shorter than the St Gotthard (*see* Fig. 19). The path of the tunnel passed under the Gaestern valley through which flows the Kander river. The depth of the tunnel was estimated at

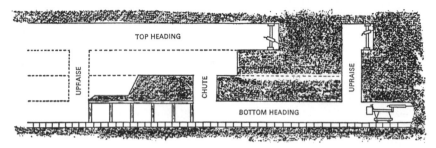

Figure 19. Driving the Loetschberg tunnel—longitudinal section.

600 feet below the floor of the valley, but it was known that the soft river-bed extended downwards for another 300 feet: this left 300 feet of rock to form the roof of the tunnel. As a precaution, the contractors commissioned another geological survey to confirm their earlier findings. It is surprising that they went to this trouble for they completely ignored its result. The geologist reported that the crevice of loose soil and water *might* reach to a depth of 600 feet; in which case the heading would run right into it. The report ended with the suggestion that a trial shaft should be sunk to confirm the actual conditions in the valley. The company paid the expert his fee and proceeded with its original plans. Two headings were worked and excavation began on the north heading in March 1907. Progress was so rapid that by the end of the year a total advance of over 9,000 feet had been made—over half the total. There was a hold-up in February of the following year when an avalanche descended upon the company's headquarters at Goppenstein. Of the thirty engineers and technicians at supper at the time, twelve were lost.

Work continued throughout the winter at a good rate and without mishap or accident until, in July, the north-bound heading was inching its way along below the Gaestern valley. Here it reached the area that had been the subject of the geologist's warning—the warning that the company had chosen to ignore. On July 24, at two-thirty in the morning, a dynamite charge was detonated. Only a thin wall of rock divided the work-face from the silt-filled crevice under the Kander. The result was instantaneous and catastrophic. The thin wall of rock disintegrated and 1,000,000 cubic feet of mud, silt and débris swept along the gallery and overwhelmed the twenty-five men who were standing back waiting for the smoke and dust to clear. The mass of sand, mud, rock and gravel was upon them before they could see it and none escaped. The tunnel was completely filled for a distance of 1,731 yards from its face. The crevice had, as the final report had suggested, reached down to over 600 feet and the loose material within it had created a 'head' of 600 feet that exerted a terrific force. As it was obviously impossible to drive through such a pressure the only alternative was accepted. The course of the tunnel was changed in mid-earth to form a detour around the crevice. At a distance of 4,675 feet in from the portal of the north gallery a wall was built, 33 feet thick, to hold back the tremendous pressure; then a new tunnel was driven off at about 4,000 feet from the mouth of the existing one. The bodies of the dead and all the equipment were left in the blocked heading, there being no way to recover them. The new heading followed the river upstream until a safer crossing was found while the south heading was diverted towards the right to meet it. The two headings met in March 1911—and the tunnel was completed well ahead of schedule despite the disaster.

Early Twentieth-century Tunnelling across the World

ALTHOUGH NOT AS formidable as the Swiss Alps, the Southern Alps on the South Island of New Zealand presented just such an obstacle to the development of railways. The idea of tunnelling was first examined in 1884, when a New Zealand Government-sponsored survey was made to construct a railway to link the east coast with the west. In spite of the considerable difficulties that presented themselves, the railway was begun—from Christchurch on the east coast and from Greymouth on the west. The problems concerned with the tunnel were many. The most suitable place to pierce the mountains, it was decided, was under the Otira pass, and, to bring the railway to the site of the tunnel, steel viaducts and bridges were built to span the many crevices and gorges. The approaches to the tunnel are 9 miles long and within that short length there are four steel viaducts (one of which is 236 feet high), three bridges, and seventeen other tunnels. In addition, there are many steep gradients.

Tunnelling New Zealand's Alps

It was not until 1907, when the two railheads had met the mountains, that work started on the Otira tunnel. The railway's path over the mountains went through Arthur's pass, a junction of two precipitous valleys on the side of the 9,000-foot-high Mount Rolleston. The 5½-mile-long tunnel was to be bored from two ends, and since these were difficult of access at the best of times and quite inaccessible during winter, living accommodation was built at each portal as well as workshops, storehouses and offices. Pneumatic drills were used, the compressors being powered by electricity obtained by harnessing waterfalls at each end of the tunnel. From the very start of operations the weather was a problem. The permanent snow-line is some 3,000 feet lower in the Alps of New Zealand than in those of Europe and the surveyors who mapped out the course were often imprisoned for days at a time in their flimsy tents. The portal installations were built in blinding snow during the winter, flooding during spring and intense heat in summer.

When tunnelling was started rotten rock and crumbling shale demanded a huge unforeseen expenditure on timbering and lining and was the cause of many accidents. Then there was a series of strikes, each time for more money, and by 1912 the contractors had had enough. The government relieved them of their undertaking and put the Public Works Department in control. In all, the work dragged on for fifteen years. The breakthrough occurred in July 1918 and the tunnel was finally completed in 1923.

The Otira tunnel enters the mountain on the east side at an altitude of 2,435 feet above sea-level and after dropping for 850 feet on a steady decline of 1 in 33 it emerges at an altitude of 1,586 feet. Over 325,000 cubic yards of rock were excavated at a cost of £1,500,000.

The East river gas tunnel

In 1910, when the Otira tunnel was running into its first difficulties, work began on a gas-tunnel under the East river in New York. Its purpose was to convey coal-gas from Long Island City to the rapidly increasing population of Manhattan. The tunnel was to be 18 feet high by 16¾ feet wide, covering the 4,662 feet between Astoria and the Bronx. To avoid the possibility of running into water it was decided to tunnel 250 feet below the river-bed, two headings being driven from opposite shafts. The work proceeded rapidly and without serious accident until the summer of 1913 when the two bores met and holed through at the upper section. All that was left undone was the excavation of the remaining rock at the bottom of the passage. After a banquet in the tunnel to celebrate the breakthrough, drilling restarted to remove the remaining rock. It was here that the Astoria–Bronx tunnel's troubles began. A batch of holes drilled in the Astoria side-heading suddenly began to gush water while, a little later, the same thing happened in the Bronx heading. It was fortunate that bulkhead doors had been installed at intervals, for the pressure of the water increased so suddenly and to such an extent that the men in both headings had to down tools and retreat rapidly. In the Bronx heading they managed to close a door as they retreated. But on the Astoria side there was no time for delay—as the men raced to the exit shaft. The water brushed aside the efforts of the pumps, flooded the tunnel and filled the shaft to the level of the river.

It was noticed that the water level in the Astoria shaft rose and fell with the tide, thus proving to the engineers that the river was the source of the flooding; pumping, therefore, would be fruitless. Floats, fitted with grouting machinery, were positioned above the estimated site of the leak and holes drilled in the river's bed-rock. Into these, cement was forced under pressure in an attempt to stop the hole. Nearly 100 tons of cement were forced into the rock, while pumps were set working in the shaft. Still the water level did not change. The drastic decision was then made to fill the whole space between the two underground bulkheads with cement by forcing it through the Bronx bulkhead which, having been closed in time, was still accessible from the shore end. The 250-foot head of water pressing down into the tunnel created a pressure of over 100 pounds per square inch, and in consequence the job of piercing the door with a 2-inch grouting pipe was one that took five days to accomplish. No less than 1,000 tons of cement were forced into the flooded heading at a pressure of 400 pounds per square inch before the water level in the Astoria shaft responded to the pumps. The cement was left to harden, additional pumps were brought up and set to work. Gradually the water level sank but when it reached the bottom of the shaft, a section of the cement grouting gave way and the shaft filled up again. Another 200 tons of cement were forced through the Bronx bulkhead before the heading could be pumped dry and on its floor were found hundreds of stranded fish. With the leak effectively stopped, work was resumed and the tunnel finished in 1916.

A tunneller's bad dream—the Tanna tunnel

If a tunnel engineer had a really bad dream it would probably be about boring through ground that presented every known tunnelling hazard plus a few new ones. The Japanese engineers who spent sixteen years burrowing under Takiji Peak lived such a nightmare, for they experienced an unparalleled series of disasters which took at least seventy lives. The Tanna railway tunnel is on a line built to reduce the journey between Tokyo and the industrial city of Kobe. It is 5 miles long and contains a double track. Work started from both ends on 1 April 1918, and a drainage tunnel was incorporated along the floor of the tunnel proper as it was expected that some water would be encountered.

On 1 April 1921—the third anniversary of the start of the work—200 feet of rock fell in suddenly, killing and burying sixteen men and trapping seventeen others at the face. Work at once started to burrow through the fallen rock to rescue the trapped men and it continued for seven days and nights before the survivors were brought out. This was the first in the list of disasters that dogged the building of the Tanna tunnel. In February 1924 an inflow of mud buried sixteen men, all of whom were suffocated. Twice the tunnelling-shield was buried by a roof collapse—once with the loss of its crew; sometimes the pressures caused by swelling clay seemed to bend the massive steel roofing beams as easily as if they were made of rubber; underground streams were encountered that turned the hard clay into flowing mud—hot springs transformed the heading into an oven, albeit a very insecure one; then, enormous pressures from the overhead rocks required concrete lining up to 6 feet thick. But worst of all was the water. No sooner was one inundation dealt with than the work ran into another. By March 1925, water was pouring into the west heading at a rate of 18,000 gallons a minute—far more than the capacity of the drainage system installed in the tunnel floor. It was decided to build another drainage tunnel, 40 feet from the main one and parallel to it. Six feet square, a major tunnelling feat in its own right, it could deal with 70,000 gallons a minute, but before it was finished the inflow of water and silt increased suddenly and without warning to 65,000 gallons a minute. This was nearly three times the amount of water met with in the Severn tunnel when the flooding was at its peak. This tremendous torrent of water put a stop to work in the main tunnel for the next year while the drainage tunnel was completed. Then the drainage tunnel itself ran into water and drifts were excavated to drain the drain. Eventually the mountain was honeycombed with tunnels; at one place as many as five parallel headings were worked.

With the water under control, work proceeded slowly and carefully in both headings and although new inflows occurred with terrible regularity they were each dealt with. Then, on 26 November 1930, nature dealt an additional blow to the men under Takiji Peak. As the mountain swayed and trembled from the shocks of an earthquake a section of the tunnel collapsed. Five men were buried by the fall and only two were dug out alive. The Tanna tunnel was eventually holed through in 1934.

The Great Apennine railway tunnel

In the square outside the railway station at Bologna stands a monument bearing the inscription '*E Tenebris Lucent*', followed by a list of names. The inscription translates 'they shine out of darkness', and the names are those of the ninety-seven men who died during the thirteen years it took to build the Great Apennine tunnel between Grizzana and Vernio. The tunnel is 11½ miles long—only 1,546 yards shorter than the great Simplon tunnel—but while the latter is the longest railway tunnel ever built, the Apennine may be considered a greater work because it was built for a double track, whereas the Simplon is two single-track tunnels. The railway line that penetrates the Apennine Mountains is known as the Bologna–Florence '*Direttissima*', an expression that has no exact counterpart in English and which means a railway line with the maximum possible number of straight stretches, curves of a large radius, no level-crossings and the smallest possible number of gradients. As may be expected, a '*direttissima*' pierces obstacles rather than circumvents them and it will, of necessity, pass through an unusually large number of tunnels; the Bologna–Florence line has thirty-one tunnels with a total length of 22½ miles, to build which 101,250,000 cubic feet of material was excavated.

In addition to the Great Apennine, two more of these tunnels were major undertakings. The first of these, the Monte Adone tunnel of 7,694 yards length, driven from both ends with no particular difficulty, was completed in five and a half years. The second was the 3,334-yard-long Pian di Setta tunnel, driven through very difficult ground, with large pockets of methane gas that slowed the advance considerably. Work was begun from the northern side in 1921 and from the south in 1925. The breakthrough was made in 1928.

The men working on the Pian di Setta tunnel may well have considered that they were having a hard time, but in fact their working conditions were envied by the men who, at the same time, were battling with all the tunneller's enemies—treacherous ground, marsh gas, seemingly endless quantities of water, explosions and fire. One after another, and sometimes in combination, these elements presented themselves to be dealt with before the advance could continue. The ground to be tunnelled was not the solid rock of the Alps, but soft clay and running sand, a combination which, as we have seen, forms about the worst possible ground for tunnelling. As it was clear that progress through such soil would be very slow, it was decided to sink two inclined shafts about midway on the tunnel's line and to drive two more work-faces outward from them. The shafts were driven at a distance of 400 feet apart at an inclination of 27°, their lengths being 1,683 feet and 1,605 feet respectively. At the bottom they were connected by a wide gallery in which the pumping machinery and other necessary equipment was installed. A cable-drawn railway was then built in each of the shafts to convey men and materials to and from the work-faces.

The very magnitude of the project necessitated a vast amount of preliminary works. To transport machinery and materials to the entrances of the tunnel, a 3 feet 1⅜ inches gauge service railway was built, comprising two lines of 17 and 13 miles respectively. The rolling stock for the lines comprised fourteen steam-locomotives and 216 trucks which, between them, shifted a total of 2,500,000 tons of material. The cost of these two lines alone was over £480,000—an amount of money which would, in these days, be expressed in millions. The entrances to the inclined access shafts were over 800 feet up on the mountain and to supply them an aerial rope-way, over 5 miles long, was erected by which some 75,000 tons of material and equipment were moved. At both portals of the tunnel and at the site of the shafts, towns were built to accommodate the thousands of workmen, engineers, office staff and their families. In addition to cottages there were schools, hospitals, maternity homes, shops, cinemas, kindergartens and churches; an aqueduct was built to bring in fresh water while an oil-burning power-plant supplied electric light.

Irruptions

Tunnelling began from both sides in 1920 and as the bore penetrated the mountain, ventilating plants were installed to supply portals and shafts with 143,000,000 cubic feet of fresh air every day. To channel this huge quantity of air, a temporary brick bulkhead was built, running longitudinally through the completed sections of the tunnel to form two air-ducts, one of which brought fresh air up to the work-face while the other returned bad air to the portal. Transport to and from the work-face was by means of compressed-air locomotives. The heading was seldom free of methane gas which, because of its low weight, rose to the roof of the tunnel to gather above the timbering. This gas, both poisonous and explosive, is liable to go off when a detonator is fired. Trained men, with special instruments, kept a constant watch for gas which, when detected in small quantities, was set alight by fuses. When large quantities of the gas were found, they were dispersed by pumping air at 1,350 cubic feet per second to dilute the gas and render it harmless. But in spite of these precautions, explosions and fires were almost commonplace occurrences that caused many injuries and delays in the work.

One large irruption of gas stopped work in the north heading for seven months. Set off by a blast, it roared through the heading and set light to the timbering for several yards. The burning timber, in its turn, ignited another large pocket of gas that ripped the ventilating bulkhead to pieces for a distance of 1,100 yards and completely destroyed the ventilating plant in the north heading. Within minutes, the heading became filled with dense, choking smoke and the temperature shot up to 212° F. In the confusion that followed, as hundreds of men made for the portal to escape the unbearable heat and choking fumes, some managed to board the little air-propelled train. But most were forced to feel their way along the sides

Plate 18. Moving a drilling 'Jumbo' up to the face of the Orange–Fish tunnel.

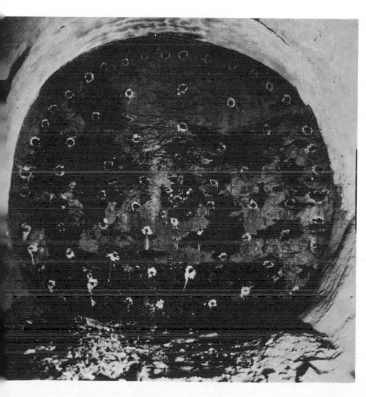

Plate 19. The drilling pattern for explosive charges, to give a 5-foot advance. The tunnel diameter is 20 feet 4 inches.

Plate 20. A hard-rock tunnelling machine.

Plate 21. Universal soft-earth tunnelling machine.

Plate 22. Inside the Orange–Fish tunnel.

Plate 23. Teebus Township. Housing for workers at the southern (outlet) end of the Orange–Fish tunnel.

Plate 24. A spectacular view of the Durch tunnel on the Bern–Loetschberg–Brig railway line in Switzerland.

of the heading, coughing and choking and, in such intense heat, feeling that the fire was upon their backs. For the men near the work-face, this ordeal was to last for over an hour as they half staggered, hardly able to breathe, through 3 miles of tunnel before they reached fresh air and light. It was a miracle that no one was killed.

At the risk of being burned or crushed to death, a party of men wearing breathing apparatus drove the compressed-air engine right up to the blaze and flooded the area with water. This checked the fire but did not put it out, so work was started to rebuild the bulkhead and install new ventilating machinery. It seems incredible that men could tolerate, let alone work in, such conditions but work they did, twenty-four hours a day, until further irruptions of gas transformed the heading into an inferno. It was then decided to wall in the blazing section and to leave the pocket of gas to burn itself out. With the burning section sealed off a by-pass tunnel was bored. Seven hundred feet long, it left the main heading at a point in front of the cross-wall, continued parallel with and 50 feet away from the original path, to rejoin the latter well beyond the burning section. Work was then continued while the gas burned itself out. When, five months later, the flames went out and the ground cooled down the heading was continued to join up with the advance that had been made beyond it.

Flooding

When not engaged in dealing with gas, fire and explosion the tunnellers were battling with water. From underground caverns and streams it poured in through bore-holes, flooded up through the floor and cascaded out of the walls of the headings in quantities that increased with the tunnel's length. To cope with this, canals were dug along the entire length of the headings and a total of twenty-seven electric pumps at the portal ends of the canals cleared them of water at a rate that sometimes reached 13,000 gallons a minute. It was found necessary to maintain this drainage system even when the tunnel was finished, as the huge inflow of water never diminished. During the five-year period ending in March 1930, 4,735,000,000 gallons of water were removed from the Great Apennine tunnel—a quantity which, it has been estimated, would fill a lake $1\frac{1}{2}$ miles long, $\frac{3}{4}$ mile wide and 33 feet deep.

At times, when the irruptions were at their worst, the flooding got the better of the pumps despite their tremendous capacity. At one point, when the pumps were working full out to keep the water within the level of the canal, a fresh incursion occurred that produced another 4,000 gallons a minute. Within a short time the heading was flooded and all work stopped. A brick wall, 10 feet thick, was built across the heading to hold back the water, then nine steel pipes of 6¾ inches in diameter were let in to the source of the water to drain it off. At the same time, seven pipes, each about 20 feet long, were driven through the wall and through them was injected a cement grout in an attempt to close up the fissures in the rock through which the water was flowing. When this failed, another attempt was made, this time using a mixture of cement, pieces of wood and scrap metal. This also failed, so the wall was demolished and the fissures were stopped by hand, against tremendous pressures by men working almost submerged in rushing water.

The advance headings of the northern section met in November 1929, and in the southern section the following month, the tunnel being finally completed in 1934. A total of 69,000,000 cubic feet of earth were excavated during the work and 16,000,000 cubic feet of lining built. Men employed underground totalled 1,300, with another 550 working in the yards. The consumption of electricity was 110,000,000 kilowatt hours and 981 tons of dynamite were used in blasting. The total cost of the Great Tunnel was £5,100,000—an average of £252 per linear yard. In spite of all the precautions that were taken to protect the lives and health of the men, every mile of the Great Apennine tunnel claimed eight lives.

13

Changing and Unchanging Aspects of Tunnelling

IN CONSIDERING THE building of major tunnels during the nineteenth century and the first of the great Alpine tunnels in particular, it is difficult to escape the conclusion that the attitudes of promoters and engineers towards their workmen had altered but little since the days of the ancients. But by the turn of the century, modern methods and machinery, as well as a more humanitarian attitude had completely changed the situation. The new attitudes are particularly apparent in the meticulous attention that was bestowed upon the welfare of the workmen engaged in building the Moffat tunnel in Colorado during the 1920s.

The tunnelling of the Rockies

The idea of penetrating the Rocky Mountains with a railway tunnel originated with David H. Moffat, builder of the Denver, Northwestern and Pacific Railroad but, because of the expense involved in drilling and blasting through 6 miles of extremely hard mountain rock, nothing came of the scheme during his lifetime. Trains crossing the Continental Divide had to climb to a height of over 11,500 feet. The ascent was not without its hazards, and over 2 miles of snowsheds were erected to protect the line from avalanches. The curves on the line were dangerous and the gradients such that special locomotives were necessary to pull trains over the mountains, sometimes as many as five of these being required to haul a single train. One 91-mile length of track took over fourteen hours to cover—an average speed of less than 7 miles an hour.

It was not until twelve years after Moffat's death that work started on the tunnel which still bears his name. In 1913 the Denver and Salt Lake Railroad approached the City of Denver with a proposal that the tunnel should be built as a joint venture between the Railroad and the City. Denver had long been considering the possibility of building a water-supply tunnel through the Great Divide and a partnership, it was suggested, would save both sides a great deal of money. The discussions to bring about this partnership dragged on for nearly ten years before complete agreement was reached, but in 1923 the first earth was finally broken. As it was thought that the work would, in many ways, be similar to that on the Simplon tunnel, it was decided to proceed in the same way, that is, a pilot tunnel would be driven simultaneously and parallel with the main tunnel, but in this case the pilot would be subsequently broken out to become the water conduit. The 6-mile-long railway tunnel was to be 16 feet wide by 24 feet high while the water conduit was to be a circular shaft, 12 feet in diameter. Driven 75 feet south of the railway tunnel (centre to centre), the water tunnel was 7½ feet above the main tunnel's permanent grade.

Camps at both portals were established early in 1923 and by the end of that year they had developed into townships. Gathered around the administration buildings at both ends were power plants, machine shops, tool-sharpening shops, canteens and recreation rooms. The sleeping huts, each containing accommodation for thirty men—two to a room—had electric lighting, laundry facilities and hot showers. For married men there were cottages set apart in small villages complete with schools and churches. There were hospitals equipped with operating theatres and X-ray machines, with a full-time staff of nurses and doctors.

The new emphasis on safety

Within the tunnel itself, safety measures were most stringent. Before each blast, all lights near the work-face were turned off by switches enclosed in locked boxes; crews were strictly forbidden to approach the face until the lights had been turned on again by the engineer in charge. The importance of this precaution was more than once demonstrated. On one occasion a shift foreman ignored the order and returned to the face before the lights were turned on. He was killed instantly when a falling rock struck him on the head.

Because of the high altitude of the works and the constant low temperatures and high winds, men were encouraged to report sick at the first sign of a cough or cold, thus combating the high risk of pneumonia. In the tunnel itself the men were compelled to wear oilskins as a protection against the continually dripping water. As a result of these precautions the sick-list, apart from casualties, was no higher than would be expected on any normal work-site.

Such was the magnitude and fame of the work and the glamour attached to it, that engineering students came from all over the United States in search of positions as drillers and muckers, thus to gain experience. The average pay was $5.15 for an eight-hour shift (equivalent in today's English money to £1·70p). Three shifts a day were worked, with each shift taking a meal into the tunnel with them to ensure uninterrupted work; unlimited supplies of hot coffee were sent up the heading by electric locomotive. Drinking was prohibited at all times and gambling was discouraged.

The water tunnel and the pilot heading for the railway tunnel were advanced simultaneously by a system of driving the two headings alternatively. The dimensions of both headings were 9 feet 8 inches, the railway heading later being enlarged to 16 by 24 feet to take a single line of track. The water tunnel was kept in constant advance with that of the railway heading for three reasons: (1) to use it as a traffic way for the removal of spoil from both headings; (2) to carry up materials and services (i.e., electricity, water and air lines) to both headings; and (3) to determine the nature of the ground to be traversed well in advance of the larger heading. The system of working the two headings alternatively ensured the full employment of the shifts. After drilling and blasting one round in the water tunnel, the drilling crew made their way through a cross-cut into the railway heading and started drilling there while the shots were placed and fired in the water tunnel. By the time this was done and the muckers had cleared away the débris, the drilling crew were ready to return to the water tunnel and start another round.

> To the miners, the firing of a round becomes a purely perfunctory matter, but to one unfamiliar with underground excavation, this event is the most dramatic moment of the cycle. Sheltered by the protecting wall of the cross-cut, to which all miners and machinery have been withdrawn, the curious visitor waits, nerves tensed in expectation. Perhaps he may wonder at the unconcern of the miners, idly chatting or bantering one another. Presently there comes a dull, low boom.
> 'Is that the blast?' is the inevitable question of the novice, voiced in a tone of half disappointment. Before a reply can come the questioner needs no answer. A pulse-like quiver follows the first report; a curious, low, beating sound that seems to shake the walls and the ground beneath the feet as tho [*sic*] a seismic disturbance has occurred. The fluttering vibration enters the human body, throbs through the veins . . . The thought occurs

involuntarily that if this terrible pulsation does not cease the solid rock itself must give way and crush the human atom beneath its incalculable weight. Another explosion follows—another—another—irregular in time, seemingly interminable in duration.[1]

It had been estimated by the surveyors that of the total length of over 6 miles only 1,500 feet would pass through soft ground that would require timbering, but this estimate soon proved to be over-optimistic. At first, small amounts of water and soft ground were encountered in both east and west headings, and these were easily dealt with. Then, in February 1925, trouble began with a considerable flow of water in the east heading. At the time of the influx the heading had not reached the apex of the tunnel and the water, which was gushing in at the rate of 1,800 gallons a minute, was diverted through a cross-cut into the railway tunnel where it followed the gradient to the east portal. The source of the deluge was traced to a lake some 1,100 feet above the workings that was leaking through a fault in the rock. The fissure was sealed with cement and the east heading continued its advance, passing the apex of the tunnel in November 1925. From then onwards, 'it seemed that every device of mechanical ingenuity, every time-saving expedient, was promptly overmatched by an intelligent and malignant force within . . . the mountain'.[2]

The hazards of nature

On 28 February 1926, a round having been fired in the east railway heading, the mucking team was coming up with its equipment. Suddenly the pile of broken rock before them started to swell and heave as if alive. Then, of an instant, great torrents of water poured into the heading, forcing drillers and muckers to turn and run, leaving drilling jumbo and mucking machine to be submerged within minutes. Soon the railway heading was full to 300 feet back from its face. Here the water flooded through a cross-cut and filled the water conduit up to the apex—a distance of over 1,000 feet. For six days the water continued at a rate of 3,100 gallons a minute, bringing with it a mass of fine silt that deposited along the floor of the tunnel for 1,200 feet. After a few days of this the flow decreased to 1,500 gallons a minute which was dealt with by two pumps.

Then, on April 21, a blizzard descended on the east-side works. Such was the fury and strength of that storm that great baulks of timber were thrown about. The men working at the portal, tough as they were, took shelter in buildings or in the mouth of the tunnel. The latter saw the

[1] E. G. McMechen, *The Moffat Tunnel* (Denver: 1927), Vol. I, p. 222.
[2] McMechen, *op. cit.*, p. 233.

power lines that supplied the pumps whipped up and snapped. The pumps in the tunnel stopped and within a few minutes the water had again reached the apex. Fortunately the men on shift at the work-face had just enough time to lift the pumps on to wagons and haul them over the apex to safety. After ten hours' work the electricians had the pumps going again, but by this time the water was flowing out of the east portal. Pumping was restarted and when the water began to recede the pumps were mounted on to wagons and advanced, still pumping, behind the retreating flood until the work-face was clear of water. Then, just as drilling was about to start again, there was another power failure. Again the water rose rapidly to the apex, this time submerging one of the pumps before it could be moved. With the aid of extra pumps the tunnel was cleared again in five days, but no advance could be made in the railway tunnel, the source of the flooding, for another three months.

In his half-yearly report on the progress of the work, the president of the Moffat Tunnel Commission wrote:

> The conditions encountered since my January report have been very grave. The water at East Portal and the soft ground conditions at West Portal have continued to delay progress and greatly increase costs. The additional cost of timbering alone occasioned by the soft rock amounts to about two million dollars. The first estimate of bad ground made by our engineers and geologists was one thousand, five hundred feet. Our present estimate is that when the tunnel is completed we will have nearly three and one-half miles of timbered tunnel.

While the engineers on the east side were struggling to hold back water, conditions on the west side were even worse. The geologists had prophesied that, apart from some isolated patches of soft ground, only solid rock would be met with on this side. The actual excavation revealed very different conditions. The seismic disturbances that had forced the mountains to the surface of the earth millions of years before had also shattered them in places so that on the western side of the mountains they were networked with faults. Water from the surface had penetrated these fissures, saturating the ground to such an extent that some of it was fluid. Even where the ground was solid it was not firm. Under the enormous weight of the mountain above even the most massive timbers shattered; steel wall-plates and 2-foot-square girders twisted and bent and gave way to the seemingly infinite pressures of the moving ground. The problem seemed insoluble until it was tackled by George Lewis, general manager to the project, who invented a machine to deal with it. Known as the Lewis Girder it was, in effect, a version of the poling-board method, consisting of an immensely strong cantilevered bar that held the wall-plates in position. Placed in a top heading so as to sustain 60 feet or so of roof, the girder allowed excavation beneath it to full size, then, when the permanent lining had been built, it was moved along to the next section. Cantilever-support bars were not a new idea, but all previous ones had to be de-molished and rebuilt for each tunnel section. The Lewis Girder moved ahead on a railway track without being dismantled. The following tech-nical description of the machine may be of interest:[1]

> The new needle bar consists of two 65 ft. steel girders, $3\frac{1}{2}$ feet in depth, tied together by rigid struts and braces and carried on dollies running along steel channels placed on 12-inch-by-12-inch stringers[2] resting on the bench, but it is designed to cantilever to the rear beyond the end of the bench a maximum distance of 20 feet, where it serves as a support for the arch timbers while a shovel removes the bench and the posts are being set. The roof load is transferred to the girders by means of cross arms, sus-pended from these girders by stirrup hangers[3] which extend beneath to the wall plates. The slack is taken up by 15-ton screw jacks. When excavation has proceeded to the maximum 20 foot overhang of the needle bar, it is moved ahead to a new position. This new needle bar carries an endless belt conveyor between the girders so that muck from the 18-foot-by-10-foot widened heading into which the machine must move for the enlarging operation, can be carried out beyond the bench and dumped thru a chute into waiting cars in the completed tunnel . . . Such attachments to the girder as electric hoists for raising posts, light wiring and piping add to its convenience and indicate its wide range of use.

Not only did the Lewis Girder overcome the difficulties that threatened to paralyse the workings in the west heading, but it actually brought down the average cost of excavation from $3.61 per cubic yard to $2.24 per cubic yard.

As the two headings grew closer together a great rivalry developed between the two teams as to which one would make the breakthrough. On the morning of 12 February 1927 the men in the east heading knew that contact was very near. Grabbing a bar of $\frac{5}{8}$ steel between them they forced it through into the west work-face. There was once, and may still be, a superstition among tunnellers that the side that fails to break through first will suffer a curse, so the west team grabbed the bar as it appeared and tried to pull it out of the hands of their rivals and through the dividing wall to their own side. A subterranean tug o' war followed as the two opposing teams each heaved and strained to wrest the bar from the hands of the

---

[1] Description given by C. A. Betts, Officer Engineer to the Moffat Tunnel Com-mission.

[2] *Stringer*, a heavy plank or timber placed in a horizontal position.

[3] *Stirrup hanger*, a drop support attached to a girder to carry the end of a joist or beam.

other. The east team, however, anticipating this battle, had bent the bar so that its end would not pass through the hole and after a long struggle they tore it from the hands of the others. The Moffat tunnel was holed through.

The White House button

It had been arranged that the dividing wall was to be demolished by a charge of dynamite fired by President Coolidge by remote control from the White House in Washington. This arrangement was the cause of great resentment—even fury—on the part of the tunnellers themselves who, after four years' work, were angry that the last, symbolic blast should be fired by someone who had never even seen the tunnel. There was more than one conspiracy to anticipate the ceremony, but a close watch on both work-faces by the authorities prevented any surreptitious or premature blasting. On the evening of February 18 the President touched a button on his desk. Twenty-four charges exploded in succession and when the smoke cleared the governor of Colorado shook hands with the mayor of Salt Lake City beneath the Rocky Mountains. The main heading was holed through in the following July, while the first train steamed through on 26 February 1928, completing in twelve minutes a journey that had previously taken seven hours.

In all, 750,000 cubic yards of rock were taken from beneath the mountains during the building of the Moffat tunnel—the equivalent of 64,000 railway wagon loads. To accomplish this, 2,500,000 pounds of dynamite were exploded, 700 miles of holes were drilled in the rock and 800,000 pounds of drill steel were used up. For timbering, 11,000,000 feet of board were used—equal to 2,000 miles of 1 × 12 inch plank. The total cost of the work was $18,000,000. Considering the magnitude of the project and the fearsome difficulties that were successfully negotiated, it is a tribute to both promoters and engineers in charge of the work that only eleven lives were lost during the whole, vast operation.

The Rove canal tunnel

The year 1928 also saw the opening of the last, and incidentally the largest, of the great canal tunnels—the Rove tunnel, making the port of Marseilles accessible to water-borne traffic of the Rhône. Cut through limestone, the tunnel is over 4½ miles long, 50 feet high and 72 feet wide—enough to allow two 1,200-ton barges to pass each other at any point. From Marseilles the tunnel carries the canal through the rock and into the lake, Etang de Berre; from there the waterway crosses the plain of Crau to meet the Rhône at Arles.

The tunnel was the result of a joint effort between the French Government and the Marseilles Chamber of Commerce. Construction started in 1911 at the south end and at the north end in 1914. Work was suspended during the First World War owing to the shortage of labour, but after a while German prisoners-of-war were put to work and it was they who actually holed through in 1916. The final work of breaking out and finishing was subject to many delays and the work took a total of seventeen years. Because of its huge dimensions, twice as much rock was excavated from the Rove tunnel as was taken from the Simplon, which is three times as long; from the time of its opening until the Goat Island tunnel was built in San Francisco, the Rove was the tunnel with the largest bore. It is by far the largest canal tunnel ever built.

14

Some Major Twentieth-century Tunnelling Achievements

THE CONSTRUCTION OF Piccadilly Circus underground station must rank as one of the most remarkable tunnelling feats in history, if only because its 155-foot-by-144-foot hall and the accompanying maze of tunnels were excavated through a single shaft only 18 feet in diameter, without interfering with the passage of the 25,000,000 people who used the old, existing station yearly and without affecting the dense road traffic overhead—even though the entire work was carried out from the central road-island. The difficulties facing the engineers were many and serious.

The design and the task

The new station was designed for a traffic of 50,000,000 passengers a year. Apart from the huge underground hall there were to be six long subway-tunnels leading to separate exits to all the important streets, a smaller hall 42 feet below the main hall, two inclined tunnels containing five escalators connecting the two halls, and two more containing six escalators leading to the two stations. In addition there were various communication passages and stairways connecting the railways. Below the Circus lay a large and complicated tangle of pipes for electricity, gas, water, telephone, telegraph and hydraulic power, and these had to be moved without interfering with the services; while for good measure a large sewer passed directly through the centre of the main hall site. First the 18-foot working-shaft was dug to a depth of 95 feet and from it headings were driven out at the required levels; the lowest heading for stairways and passages connecting the Bakerloo with the Piccadilly railways; the middle for the lower hall and the tunnels for the lower escalators; the highest for the booking hall and the main escalators. As the various levels honeycombed outward from the shaft the excavated material became too much to handle through the working shaft, so the surplus was taken through a temporary shaft to a specially built platform on the Bakerloo line. Here it waited until the passenger trains stopped for the night, when it was loaded on to trucks and taken away.

To accommodate the displaced service-pipes a special tunnel was built, slightly larger than the railway tunnels and 550 feet long. To excavate the main hall and to place its roof in position a number of shallow headings were driven out to a distance of 400 feet from the central point. As they advanced, steel roofing joists were positioned, supported by timbers. At the end of four of the headings, shafts were sunk and in these the foundations of the four supporting pillars were built. With the columns in position the main roof girders were placed on them and the temporary timbering removed. While this work was proceeding a circular bottom heading was driven around the entire perimeter of the booking hall, and within this was built the foundation for the wall of the hall; the wall was then built up to the roof of the heading. This done, a second circular heading was driven above the first and the wall built up further to be finally completed through the headings made for the roof. With wall and roof in position there was now no danger of the ground falling in and a whole network of tunnels was then driven beneath the roof. When they were completed, the walls dividing them were removed. The result was the 155-foot-long and 9-foot-high oval hall from which were driven the two main escalator tunnels. Work started on the station in February 1925 and it was opened to the public in December 1928.

Tunnels and the Motor Age

When canals gave way to railways, the horse and the 'legger' gave way to the steam-locomotive, thus creating a new problem for the tunnel engineer —that of ventilation. Except in extremely long tunnels or those passing under water, the difficulty was overcome simply by placing ventilating shafts at intervals along the tunnel's route and these were usually adapted from the various working-shafts used to dig the tunnel. These shafts, acting in conjunction with the displacement of air caused by the moving train, proved perfectly adequate even in very long tunnels. In tunnels where it was not possible to supply ventilation shafts, as, for instance, under mountains or rivers, conditions for the traveller could be very bad indeed. In his book *Son of Oscar Wilde*, Vivian Holland described a journey through the St Gotthard tunnel in 1896:

> We had been warned that it would be an awe-inspiring experience, so we were prepared for something peculiar. Electric trains had not yet made their appearance and, as there was no ventilation along the whole twelve miles of the tunnel, extraordinary precautions had to be taken to prevent people being suffocated by fumes from the engine. The engine-driver and the fireman wore gas-masks, and before the train entered the tunnel the guard went right through the train to make sure that all windows and ventilators were properly shut. An open window or ventilator would almost certainly have had a disastrous effect on the passengers in that particular compartment.
>
> The train took half an hour to pass through the tunnel. The smoke gradually filtered through the edges of the windows into our compartment until we could hardly see across it; the solitary oil-lamp in the roof grew dimmer and dimmer, and the temperature kept on rising. To add to our discomfort, the noise of puffing and clanking was deafening. Presently, above all the din, some small children in the next compartment got into a panic and started screaming. And just as we thought we could stand it no longer and wanted to open the carriage door and throw ourselves out, the blessed daylight reappeared; we were out of the tunnel and the train stopped. The carriage windows were so blackened by smoke that we could see nothing at all through them; so we got out on the line and walked about while the carriage aired and men sluiced the windows down with mops. We felt thankful to be alive; and a little surprised too. Nowadays, a spotlessly clean electric train does the journey in twenty minutes and in comparative silence.

In some cases the engineer could rely upon a natural air-flow to carry away the fumes from a steam locomotive. This can be created in a tunnel in three different ways, operating separately or together, in the same direction or in opposite directions: by (1) the difference in temperature between the inside and the outside of the tunnel; (2) the difference of atmospheric pressure between the two portals; and (3) the passage of wind through the tunnel.

Where these conditions did not exist, fans were sometimes installed to inject air through the tunnel. With the coming of the 'Motor Age', new methods of tunnel ventilation had to be devised. With a steady stream of vehicles passing through it, the tunnel would soon become lethal if the fumes were simply drawn or forced along its section. It is possible to survive for some time while breathing the fumes from a steam engine, but not when inhaling carbon monoxide—especially when driving a car.

The first long tunnel planned especially for heavy motor traffic was the

Holland tunnel that crosses beneath the Hudson river to join New York with New Jersey. The tunnel was first authorised in 1919 and the two states agreed to finance the project as a joint venture, that, on completion, would be maintained by a toll. Clifford M. Holland was appointed Chief Engineer and by the end of 1920 he had completed his plans. The problem of dealing with large quantities of exhaust fumes was a new one in those days, and Holland studied it very carefully. Before he could design a ventilating system, it was necessary for him to know what quantities of gas were to be dealt with, so he started by making a comprehensive series of tests on motor vehicles in closed chambers as well as on the road, and accurately measured and analysed the exhaust fumes. Next, to find out the effect of this gas upon people exposed to it, he carried out another series of tests on volunteer subjects. Then, aided by his staff, he devised the method of ventilation that became known as the 'vertical transverse flow' system. Ventilation buildings at either end of the tunnel draw in fresh air through side louvres and pump it through large air-ducts beneath the floor of the tunnel. From these it is led, through flues placed at intervals throughout the tunnel, into 'expansion chambers' that run the full length of the tunnel and at each side of the road at curb level. From these chambers issues a steady, gentle current of air that mixes with the exhaust gases of passing traffic and is sucked upwards through grilled openings in the ceiling by powerful suction fans in the ventilation buildings; from there it is discharged into the open air. The advantage of the system is that there is neither wind in the tunnel nor any smell of exhaust fumes.

Work started in October 1920 from Canal Street, New York, and before long five shields were operating on twin headings and in the various drifts connecting them. The shields worked under air pressure and the inevitable blows were encountered. Holland hardly ever left the work for nearly four years—both eating and sleeping on the job. Eventually his health broke and he died two days before the first heading was broken through. The Holland tunnel is, in fact, two tunnels, each 8,463 feet long, spaced about 60 feet apart—one for west-bound and one for east-bound traffic. Completed in 1927, it was opened in November of that year when 20,000 people walked through in the wake of the official procession. It was then permanently closed to pedestrians and opened to motor traffic. Originally called the 'Hudson River Vehicular Tunnel' it was later given the official title of the 'Holland Tunnel' as a memorial to its builder.

The Mersey road tunnel

The story of another great motor tunnel, the first Mersey tunnel, starts in 1922 when a committee was appointed to consider the best method of joining the city of Liverpool to the town of Birkenhead which lie on opposite sides of the Mersey. A tunnel having been decided upon, Parliamentary approval was secured and work began in December 1925 on a main tunnel 3,751 yards long, with various branches leading to the docks. The total length was 5,064 yards and the width, which provided four lanes of roadway, was 36 feet in the main tunnel and 19 feet in the branches.

The River Mersey is a deep one, and to have tunnelled through the mud of its bed would have required a pressure of 73 pounds per square inch—far more than a man can survive, let alone work in. It was decided, therefore, to drive the tunnel through the solid sandstone that lay below the mud, and this involved driving the tunnel at a depth of 146 feet below high water.

On both sides of the river, 200-feet-deep shafts were sunk and lined with cast-iron until the sandstone was reached, after which no lining was needed. From the bottom of these shafts two pilot headings, 12 feet high by 15 feet wide, were driven from each side, one being a top heading, the other a bottom heading. As was expected, water was encountered from the start and to check it deep holes were drilled in the work-face through which cement was forced to close up the fissures in the rock in advance of the heading. Although this operation considerably decreased the flow, it by no means stopped it, and water continued to flow into the Liverpool headings at 2,400 gallons a minute.

In order to cope with the large quantities of water expected, the working-shafts had been sunk considerably lower than the depth of the tunnel. From the foot of the Liverpool shaft a drainage tunnel was driven below and parallel to the two main headings and at a gradient of 1 in 500. When this drain had advanced beyond the main tunnel, the three headings, top, bottom and drainage, were connected by vertical shafts so that the water found its way into the lower, inclined, heading and flowed to the foot of the shaft. Here it was dealt with by pumps.

On the Birkenhead side, comparatively small quantities of water made their appearance and these were dealt with by pumps. But when, after twenty-seven months work, the two sides met, all water was directed into the Liverpool drainage tunnel to be pumped up from the bottom of the Liverpool shaft.

Rock weighing a total of 1,200,000 tons was excavated from the Mersey tunnel and it was all moved by lorries during the night. With the top and bottom headings complete, work started on breaking out to full size. First, inclined shafts or chutes were cut between the top and bottom headings, through which the broken rock was loaded on to wagons running on an electric railway below.

The lining of the main tunnel, weighing over 82,000 tons, is made up of over 1,000,000 iron segments, while the annular space between lining and rock is filled with small pieces of stone, grouted under pressure through holes pre-cast in the segments. In construction, these holes were filled in with bolts and all joins between the segments were sealed with lead wire.

Following the principle of the Holland tunnel, the Mersey tunnel is ventilated by air ducts below the carriageway, and connected to the tunnel itself by openings at road level. Six ventilating stations, three on either bank, supply fresh air to the air-ducts and draw off the exhaust fumes along the roof of the tunnel. The air is extracted and replaced at a rate of 2,500,000 cubic feet a minute. The tunnel, which was opened in July 1934, took nearly nine years to complete.

A Californian tunnel

The 6½-mile-long Tecolote tunnel carries water from the Cachuma reservoir under the Santa Ynez Mountains and discharges it into the conduit in Glen Annie Canyon, California. Its builders have claimed with some justification that it is the most difficult tunnel ever driven in the Americas. For the first few months after the start of work in 1950, the cycle of drilling, blasting and mucking proceeded without incident or accident, the little water that was encountered being easily dealt with by the pumps. Then, in January 1951, when 8,940 feet had been driven, the tunnel was suddenly transformed into what one of the workmen called 'a snarling, viscious, fighting tiger' as a cascade of water suddenly erupted through the tunnel face, nearly drowning the drilling-crew who downed tools and ran for their lives. Almost as soon as the men had taken to their heels, a violent explosion of methane occurred at the work-face, hurling a blast of intensely hot air after the retreating work-gang, seriously burning the faces of eleven of them and severely singeing the rest. Rapidly 200 feet of the heading then filled with sand and water. The next few months were taken up with building heavy bulkheads to hold back the inundation and in stopping the source of it with high-pressure grouting. Then the advance was continued through the hard, hazardous, broken ground until, at last, the troublesome zone was left behind and the inlet portal heading reached a point 14,919 feet inside the mountain.

Because it was feared that rising water in the reservoir might flood through the inlet portal, tunnelling from this end was stopped and the lining was started. Excavation continued in the outlet section and made good progress until water again brought proceedings to a standstill. All attempts to grout the 3,500 gallons a minute that had to be dealt with failed. Finally, the engineers resorted to the time-honoured method of solving the problem: smaller drifts were driven on each side and ahead of the main tunnel to relieve the water pressure. Because of the large quantities of water in this section the tunnel was broken out and lined with concrete without delay. Under the floor of the heading, and buried by the concrete lining, was set a perforated pipe connected to a discharge pipe. Water accumulating behind the concrete lining was forced by its own pressure into the pipe and out of the portal.

Bathtubs

As the tunnel pushed farther into the mountain, so the temperature within it increased. At 14,000 feet it had reached 112° F, and the flow of hot water created serious difficulties in loading the explosive charges into the bore-holes; many extra holes had to be drilled for each round to allow for abandoning one or more if the water interfered with loading operations. In some cases the hot water softened and even dissolved the cartridges. Water and heat combined together in the heading to produce an atmosphere that resembled a Turkish bath, and, to keep the men cool for at least some of the time, they were taken to and from the work-face in wheeled 'bathtubs'—these were mining skips filled to the brim with cold water. The men rode from the portal to their working places up to their necks in water and, during the necessarily frequent rest periods, they again immersed themselves in the skips that were left for them in nearby sidings. The fact that the men were clothed when 'bathing' did not matter, for they were constantly saturated with tunnel water or soaked with perspiration. These conditions, though, were too much for the older tunnellers, and only the younger men could stand the hot, wet slog. These men, unfortunately, were too young to have acquired tunnelling skills. By July 1953 the contractors had to give in and the job was shut down.

With all the other units of the project completed it was vital to complete the Tecolote tunnel. Of the total length of 33,500 feet less than a mile remained between the two headings, but the thought of what troubles lay within that mile was enough to discourage even the most work-hungry contractor. Eventually, arrangements were made with two companies to finish the job as a joint venture—but at a price. The original firm had undertaken the work at a figure that worked out at $44 a cubic yard. The newcomers demanded, and got, the staggering rate of $447 a cubic yard, over ten times the rate awarded to their predecessors. The figure, however, was justifiable for it covered every possible contingency that might arise during that last mile of tunnelling.

Work restarted in January 1954 in the same dreadful atmosphere of 112° F, and with water pouring and spurting in at nearly 10,000 gallons a minute at a temperature of 117° F. Then, as if the mountain was adding insult to injury, great pockets of the evil-smelling gas, sulphurated hydrogen, seeped through the seams to foul the air and burn the eyes of the men. But in spite of all, the first four weeks showed an advance of 375 feet and thereafter a weekly average of 115 feet was attained. On 15 January 1955, the outlet portal was holed through to the inlet portal amidst great relief on the part of the engineers and rejoicing on the part of the men, as a great holing-through banquet was given in nearby Santa Barbara—who is incidentally the patron saint of tunnelling.

The Great St Bernard tunnel

The construction of the last of the great Alpine railway tunnels coincided with the decline of the railways as a means of transport and the beginning of the age of the motor-car; and yet seventy-seven years after the invention of the internal combustion engine, 35,000 motor vehicles a year were being taken by train through the Simplon tunnel alone—the nearby St Bernard pass only being open for three months of the year. In adverse conditions a journey through the pass is a formidable one. Says the guide-book: 'A difficult ascent from either side. Narrow and gravel-surfaced for 8 kilometres from the summit in both directions. Not recommended for caravans.'

The idea of a tunnel in this region to link the Swiss town of Martigny with Aosta in Italy dates from 1850 and it was revived and shelved at regular intervals. It was not until 1957 that a decisive move was made. In that year a syndicate was formed between the Swiss canton of Valais and the town of Lausanne on the Swiss side and, on the other side, the provinces of Aosta and Turin together with the Fiat Motor Company. The latter had long regarded an all weather road-route between Italy and Switzerland as an investment in Italian industry, and they put up half the Italian contribution to the project. In 1958 work started on two access roads to the southern portal and work on the tunnel itself began the following year.

To relate in detail all the difficulties and dangers that were met with during the four-year operation would be but a repetition of those encountered in the Simplon. The same round-the-clock shifts were worked —drilling, blasting, mucking, drilling, blasting, mucking. No one could be quite sure of what the next blast would bring about. Sometimes the rock, released from the tremendous pressures of the mountain, would burst like a bomb. Sometimes great lumps of granite would break off the walls or roof, seemingly of their own accord, and fly horizontally across the confined space of the heading. In some places the granite crumbled like rotten wood and had to be held in place by a system of steel pins and wire netting. But the work never ceased, although much of it was carried out by men knee-deep in hot water or mud. On the mountain, avalanches, floods and violent storms isolated the works time and again, but still four shifts a day were worked, making a steady progress of one mile a year. A total of five men died on the French side, twelve on the Italian and 800 men were injured.

The tunnel was completed in 1962 at a cost of £12,000,000. At a height of 6,000 feet above sea-level, the 3½-mile-long Great St Bernard tunnel shortens the distance across the Alps by 6 miles. It reflects sadly upon modern tourists that even under the most ideal of conditions the great majority of them prefer to use the tunnel rather than the chance of driving through 9 miles of some of the most majestic mountain scenery in the world.

While the Great St Bernard tunnel was under construction another sub-mountainous passageway was being driven through the Alps. This tunnel pierces the massive Mont Blanc itself and is a project that dates from 1907. Originally a railway tunnel was envisaged but the scheme was shelved on the outbreak of the First World War. In 1928 it was revived by the French engineer, Arnold Monod, in the form of a road tunnel. A Franco–Italian–Swiss syndicate was formed in 1935 but the Abyssinian war prevented further progress.

It was not until 1953 that agreement was reached between the three governments concerned and it was then decided that the tunnel would connect Chamonix in France to the head of the Aosta Valley in Italy, a distance of just over 7 miles that would shorten the Paris–Turin and Paris–Milan routes by 137 miles and 195 miles respectively. The tunnel was commenced in January 1959, and from the start the conditions on the Italian side were far more severe than those on the French. The same difficulties were met with on both sides that were being experienced in the St Bernard project just 20 miles away. Excavation was carried out in three shifts a day by teams of forty-five men under a foreman. Each shift carried out a complete working cycle: drilling, which took two hours; charging and firing, which took one hour (plus a twenty-minute period to allow the smoke and dust to clear) and mucking-out which took just over four hours. Fifteen drills were used and, for every advance of 13 feet, 130 holes were drilled each about 14 feet deep. The problem of 'spalling', that is, pieces of rock breaking off under tension, had to be specially dealt with. This type of rock failure can take place any time within a matter of a few hours to several weeks after blasting so, as soon as possible after each blast, deep holes were drilled around the newly excavated section and into these were driven special 'roof bolts'. At first the headings on both sides were driven full section, but after advancing for a third of a mile the Italians ran into such problems of rock pressure and flooding that they were forced to resort to driving a top heading, which they later opened out to a half section as in the English method; this method was continued until the gallery reached hard granite, at which point it was extended to full section before the advance was continued.

The underground toll of Mont Blanc

The deeper the mountain was pierced, the more violent became the results of the rock pressures. On one occasion this pressure shot a drill out of its hole like an arrow which impaled the body of one of the workmen and killed him. Spalling became so severe that at one time up to forty-one hours had to be spent after each shot-firing on mucking-out and securing the rock-face. On the night of 5 April 1962, the mountain struck from a different direction. Three avalanches, one on top of the other, roared down the mountain on to the work camp on the Italian side, killing three men and injuring thirty others.

In all, twenty-three men died on or under Mont Blanc during the four years it took to complete the tunnel. On 17 August 1962 the Italians fired their last round and broke through to the French heading. The Mont Blanc Road tunnel—at present the longest in the world—demonstrates that with all the mechanical and scientific aids that are now employed, tunnelling under mountains is still very much the tough and dangerous task that it was in early days and that muscle and tenacity are its prerequisites.

15

The Channel Tunnel:
Proposals and Plans

I N 1832 A French engineer and geologist, Thomé de Gamond, started
on a survey that was to take him thirty-five years. The survey was for
a tunnel beneath the English Channel to join France with England.
Having made a number of marine soundings and gained an accurate
indication of the contours of the bed of the Channel between Dover and
Calais, de Gamond next proceeded upon a geological survey. His method
was simplicity itself. De Gamond's diving equipment consisted of two
pieces of lint fastened over his ears as a defence against water pressure, a
satchel full of stones over his shoulders for ballast, and some air-filled
pigs' bladders around his waist. Thus equipped and secured to a rowing-
boat by a long line, de Gamond was rowed to various points of the
Channel where he jumped overboard to be quickly weighed down to the
sea-bed by his bag of stones. On arrival he scooped up a sample of the
channel-bed, discarded the satchel and pulled on his life line as a signal to
his crew to pull him up. De Gamond made hundreds of descents in this
manner, seemingly never affected with the bends although he often
reached depths of over 100 feet.

The indefatigable Frenchman

Quite early on in his investigations the French engineer decided that a
tunnel was *not* the most practical way of joining the two countries and
came up with a number of alternative methods. First he suggested the
construction of a great causeway linking Dover to Cap Blanc Nez with
three openings, each spanned by a swing bridge to allow for the passage of

shipping. Next he designed a high, massive bridge of granite and iron stretching from Dover to Calais, but later decided that such a bridge would not stand up to storm and tempest. He then abandoned the idea of a permanent link altogether and drew up a vast ferryboat, built of concrete and driven by paddle engines, that would have dwarfed those of the Great Eastern. This boat was to carry four long railway trains as well as one hundred horse-drawn vehicles.

Tiring of that idea, the unpredictable engineer collaborated with an English colleague in the design of a rail link that would have done justice to Mr Heath-Robinson. This was to be a great iron gantry running on a railway track of 100-foot gauge laid on the bed of the Channel. The top of the gantry was to rise 50 feet above high-water level and was to carry four trains. Steam engines on opposite sides were to cable-haul the gantry across the Channel. It is possible that when de Gamond saw the latter scheme on paper it was too much even for him to stomach, for his next idea was to lay an iron tube across the Channel bed after smoothing the sand with rakes and compressing it with rams—both rams and rakes to be worked by steam engines from ships on the surface. To reach the bottom of the sea at some points in the tube's route, the rakes and rams would have needed to be over 200 feet long—this realisation led to the abandonment of this scheme. Finally, de Gamond proposed to tunnel under the Channel bed, thus arriving, after a circuitous route that had taken up over twenty years, back at his starting point.

In 1856 he finished his plans for the tunnel. It was to carry twin railway tracks between Eastwear Point and Cap Gris Nez. Work was to proceed from the two ends, and both ways from the Varne, an underwater shelf of rock, 10 miles off Folkestone. At high water, the top of the Varne is only 12 feet below the surface and de Gamond proposed that a coffer-dam could be built on it, within which a shaft would be sunk to grade, 150 feet below the sea-bed. This shaft, 350 feet in diameter, was to open out into a large sub-aqueous railway station, where the passengers after the exhausting 10-mile journey from Folkestone could stop for refreshments and a rest, or climb to the top of the coffer-dam to enjoy a view of the channel and the sea air.

For ventilation, de Gamond proposed a series of built-up islands along the path of his tunnel, each shafted to allow the smoke and steam of the locomotives to escape. Later, to save the enormous expense that these shafts would involve, he suggested that smokeless engines could be designed to draw the trains. Dispensing with fires, these engines were to be powered by high-pressure steam stored in cylinders! It is strange that de Gamond, obviously an accomplished engineer, should have entertained so many crack-brained ideas. Why did he not realise, for instance, that a furnace would be required to keep steam dry long enough to carry a heavy train the length of the tunnel, or that to carry sufficient steam the containers would have to be vast in size and so massive of construction as to make them impracticable?

Political aspects and public reaction

Nevertheless, the tunnel was feasible if the engines were not and a scale-model of the project created great enthusiasm on both sides of the Channel. But continual political bickering between France and England ensured that nothing came of de Gamond's scheme and he died in poverty in 1876.

In 1867 the British engineer William Low had obtained an interview with Emperor Napoleon III and discussed with him the possibilities of a channel tunnel. Low was a thoroughly experienced tunnel engineer, having been employed as assistant to Brunel on the Great Western Railway. Later, he had set up as a consultant colliery engineer, thereby extending his knowledge of tunnelling. The Emperor's reaction to the proposal was such that Low at once formed a company and began negotiations with the French and British Governments to obtain the concession for building it. Other tunnel engineers were associated with the project, including Sir John Hawkshaw of Severn tunnel fame. Discussions and negotiations continued until 1870 when the Franco-Prussian War put an end to the scheme.

Low revived the project in 1872. The French were strongly in favour, while the British Government went so far as saying that it did not object, providing that the tunnel would be built entirely by private capital and that the Government would not be asked to make any guarantees. On this basis the Channel Tunnel Company was formed in London, with Low as engineer. A similar company was formed on the French side and concessions were granted by the respective Governments to operate the tunnel for ninety-nine years.

In 1875 a rival company was promoted on the English side under the chairmanship of Sir Edward Watkin, a railway promoter. Called the Submarine Continental Railway Company, it promptly abducted William Low to its side, came to an understanding with the French Company, and gobbled up the original Channel Tunnel Company almost before the latter's directors knew what was happening.

In 1880 the new company sunk an exploratory shaft to a depth of 75 feet at Abbots Cliff, and from its bottom a 7-foot diameter pilot passage was driven out to explore the terrain. This was found to consist of grey chalk—a good medium for tunnelling. Another shaft was sunk from the top of Shakespeare Cliff west of Dover, and from the base of this the Channel tunnel was at last started.

The intention of the company was first to drive two parallel 7-foot-diameter pilot tunnels below the sea-bed. These would be advanced from both sides of the channel and would be 23 miles in length. This, it was estimated, would take between three and four years and, when completed, the pilot passages would be enlarged to 14 feet in diameter. The engines drawing the trains through the tunnel were to be powered by compressed air carried in cylinders in special tenders, the exhaust of which would serve as ventilation. At each end of the tunnel there were to be underground stations; here the carriages were to be uncoupled from the engines and raised to the surface on a lift. At ground level steam-locomotives would haul them to London or to Paris.

One of the men enlisted to the Board of Directors was Colonel Frederick Beaumont, an expert in military tunnelling, who had invented a tunnelling-machine. This was a cylinder, 33 feet long from borer to tail, that burrowed by means of short steel cutters fixed in two revolving arms, seven cutters to each. The upper portion of the frame, in which the boring apparatus was fixed, moved forward $\frac{5}{16}$ inch with each complete revolution of the cutters. In this way a thin paring was taken from the entire chalkface of the heading making a tunnel 7 feet in diameter.

This rate of progress may well have been slow but it did have the advantage of being steady; worked by compressed air it advanced at the rate of 100 yards a week. Later, Colonel Beaumont improved his machine and doubled the rate of advance but not before the channel tunnel was abandoned.

On the occasion of a visit to the works by de Lesseps, *The Times* of 3 July 1882 gave a good description of the works under the English Channel:

> M. de Lesseps and a party of French engineers and scientific men paid a visit on Saturday to the works of the Submarine Continental Railway Company at Dover and inspected the tunnel, which has been run out under the Channel for a distance of over a mile. The French visitors, who had landed at Dover on the previous day, reached the base of operations at the western side of Shakespeare's Cliff at about 10.00 in the morning . . . the visitors were lowered six at a time in an iron 'skip' down the shaft into the tunnel. At the bottom of this shaft, 163 ft. below the surface of the ground, the mouth of the tunnel was reached, and the visitors took their seats on small tramcars that were drawn by workmen. So evenly had the boring machine done its work that we seemed to be looking along a great tube with a slightly downward set, and as the glowing electric lamps, placed alternately on either side of the way, showed fainter and fainter in the far distance, the tunnel, for anything one could tell from appearances, might have had its outlet in France. A journey of some 17 minutes, however, not counting a stoppage for refreshments when 1,000 yards had been traversed, the work-

men drawing the cars on the down grade at a fast walk, brought the party to the end of the boring—1,900 yards from the shaft and about 150 ft. below the sea.

That there would be opposition to a road link with the Continent from sections of the insular-minded British public was inevitable, and it was equally inevitable as to which side *The Times* would join. Said a leader-writer in June 1881:

The narrow seas will be no longer an obstacle to a free uninterrupted communication by land between the two countries. . . . To be able to get from London to Paris and back again without danger of seasickness by the way is an agreeable prospect for a good many travellers on both sides of the Channel. But there are others to be considered. The advantages Sir Edward Watkin offers have an obvious and essential drawback. . . . His new path for commerce and for peaceful travellers will be open in the event of war. . . . Will it be possible for us so to guard the English end of the passage that it can never fall into any other hands than our own? . . . a force of some thousands of men secretly concentrated in a Channel port and suddenly landed on the coast of Kent might seize the English end of the tunnel, might entrench itself there, and might hold possession of it for a time. A few hours might be enough; a few days certainly would. The tunnel with both ends in hostile hands, would be a safe passage ready made for the invader. Men and materials could be sent through it. The force which had possession of our end might be joined speedily enough by the whole disposable army from the opposite shore. In four-and-twenty hours it might be very largely reinforced. In three days time it would have swelled to numbers far in excess of our own. . . . If Sir Edward Watkin's tunnel is ever finished, and long before the shareholders have begun to reap the golden harvest it is to bring them, we shall hear from more quarters than one of the uses to which the tunnel can be put. A design for the invasion of England and a general plan of the campaign will be subjects on which every cadet in a German military school will be invited to display his powers. After all, what is to be set off for the alarms the tunnel must bring upon us? The journey from London to Paris is to be shortened and made more comfortable. We admit the advantage, but it can be obtained at much less cost. If the French would do as much for Calais as we have done for Dover, if they would make a proper port of it, accessible at all tides to large vessels, the means of communication between the two countries would be quite as easy as we wish them to be. A steamer of the class which runs between Dublin and Holyhead would make the passage from Dover to Calais in little more than an hour, she would be steady in all weather, and regular to her time at all states of the tide. If persons are not contented with this, if they must have the tunnel and nothing but the tunnel, we must bear in mind the price at which the whim can be gratified. It is not Sir Edward Watkin and the shareholders who have the only right to a voice here.

The British public took this as an invitation to express their right to a voice on the subject, and *The Times* became the main forum of both pro- and anti-Channel tunnels. Even the French joined in with the complaint that the British fears worked both ways—to be answered, quite rightly, by British military authorities that that had nothing to do with it. The pros, led by Sir Edward Watkin himself, asserted that the portal on the English side could be demolished immediately by 'a pound of dynamite or a keg of gunpowder', and the company even offered to construct two forts to cover the English end night and day with a battery of field guns.

The French were so keen on the tunnel that they offered to build the rail approach to their portal in the form of an elevated ramp that would sweep, in a huge circle, out to sea and back again; then in the event of Anglo–French hostilities, the ramp could be destroyed by British cannon firing from the Dover cliffs.

One proposal to make the tunnel impassable was reported by *The Times* in May 1882:

> Dr Siemens has, by request, presented to the Military Committee on the subject of the Channel tunnel an ingenious and novel plan for defending the tunnel, if constructed, from hostile invasion. He proposes that immediately above the lateral drainage there should be a driftway or tube, terminating on the tunnel side in a double arch, with numerous perforations into the tunnel, and on the land side in several chambers of wrought iron sunk into the ground. These chambers he proposes to fill with lumps of common chalk and to connect each of them by means of a pipe with a large cistern filled with dilute muriatic acid. Upon opening the communication this acid would flow into the upper portion of one of the chambers, where it would be distributed by perforated pipes over the whole area. The result of such an inflow would be a powerful chemical reaction, giving rise to a generation of carbonic acid gas, which would for half a mile or more form an insuperable barrier to the passage of human beings through the tunnel. The valves by which the acid was turned upon the chalk might be worked from a safe distance by electricity. The scheme thus briefly sketched is recommended by Dr Siemens as being comparatively cheap and easy of adoption whilst leaving the tunnel intact and fit for use after a reasonable interval for proper measures to clean it from the carbonic acid gas.

This brought an immediate reply from an anti-tunnel reader:

> Coming down here today on a part of the line where our speed was about 30 miles an hour, I took out my watch and determined how long I could hold my breath without inhalation. By emptying my lungs very thoroughly, and then charging them very fully, I brought the time up to nearly a minute and a half. In this interval I might have been urged through more than half a mile of carbonic gas without injury.

In April 1882 the anti-tunnellers presented a petition to Parliament calling upon it to put a stop to the project forthwith. The signatories included such influential names as Alfred Lord Tennyson, Robert Browning, Cardinals Newman and Manning, the Archbishop of York and Professor T. H. Huxley. Most influential among the tunnel opponents was the military hero and favourite of the British people, Sir Garnet Wolseley. His fears went even further than those of the earlier alarmists for he warned the country that

> a couple of thousand armed men might easily come through the tunnel in a train at night, avoiding all suspicion by being dressed as ordinary passengers, and the first thing we should know of it would be by finding the fort at our end of the tunnel, together with its telegraph office, and all the electrical arrangements, wires, batteries, etc., intended for the destruction of the tunnel, in the hands of an enemy. . . . Twenty thousand infantry could . . . be easily dispatched in 20 trains and . . . that force could be poured into Dover in four hours. . . . The invasion of England could not be attempted by 5,000 men, but half that number, ably led by a daring dashing young commander might, I feel, some dark night, easily make themselves masters of the works at our end of the tunnel, and then England would be at the mercy of the invader.

It is easy to smile at Sir Garnet's nervousness over the channel-tunnel proposal but it should be remembered that England, then the greatest world power in history, was entirely dependent on her watery insularity and her Navy. For eight hundred years Britain had been involving herself in the world's quarrels, but had been in a position to choose the fields of her battles and not to have them forced upon her by an invading army. The British people of the 1880s remembered still that in the days of Napoleon III it was said that he was the only person in France who was opposed to an invasion of England.

The proposal today

The successes of the German parachute corps in 1938–40 demonstrated that Wolseley might have had a point when he expressed his fears about the capabilities of a small force, 'ably led by a daring, dashing young commander'. At any rate the risk was there and there was something to be said for Sir Garnet's view that 'John Bull will not endanger his birthright, his property, in fact all that man can hold most dear . . . simply in order that men and women may cross to and fro between England and France without running the risk of seasickness'. In the event John Bull did no such thing, for in 1882 the Board of Trade stepped in and forbade any further work on the Channel tunnel on the grounds that it passed through Crown property, that is, the foreshore and the sea-bed to the line of the 3-mile limit. There the matter rests to this day despite the support for the tunnel given by Gladstone in 1887 and by a series of governmental enquiries that stretches into modern times.

Britain's coming participation in Europe's Common Market will result in a vast increase in the cross-Channel traffic of both goods and people, and the result of this would appear to be a direct link of some sort between Great Britain and the Continent. Will it be a bridge or a tunnel? And if the latter, will it be a tube lying on the ocean bed or a tunnel proper? Surely it must be one or the other, for the military objections are no longer valid. But the insular mind and the innate xenophobia of the British people will not be put off by such arguments. The current dread of the British in the nineteenth century was invasion across the channel. The enemy today is 'pollution', and 'conservationists' are up in arms, as the *Evening Standard* recently reported, to 'prevent the Garden of England being turned into a European corridor'. The most active of today's anti-tunnellers have formed the Channel Tunnel Opposition Association, the members of which include farmers who want to preserve their land and seamen who want to preserve their jobs!

The tunnel was attacked from a different quarter on 16 November 1971 when Mr Michael Clark Hutchison, Conservative M.P. for Edinburgh South, told the House of Commons that the proposed tunnel is an 'out-of-date, expensive and ludicrous conception' in these days of mobile air-craft, hovercraft and 'roll-on-and-off' ships. And so the arguments flow back and forth as they did in the 1830s and until they are permanently resolved the Channel tunnel must remain an even Greater Bore than the Hoosac.

16

The Present and Future of Tunnelling

I N COMMON WITH all other branches of engineering, that of tunnelling has, over the last one hundred years, advanced so considerably and so rapidly that tunnelling feats impossible to one generation become commonplace to the next. With the introduction of new techniques, new machines and improved materials, many of the old tunnelling skills—particularly the art of timbering—are becoming lost and this is to be regretted. On the credit side, however, the machine has improved working conditions beyond measure and reduced the number of casualties to a fraction of what it once was.

Great savings of life and limb as well as money have resulted, for example, from the use of the extensometer in rock tunnelling. Bore-holes of about 30 feet depth are drilled into the roof, walls and floor of the pilot heading at intervals of about 200 feet. Into each of these holes is placed an extensometer—an instrument that measures the degree of tension or compression existing in the ground to be tunnelled. By knowing in advance the stresses and strains to be met with, the engineer can calculate the strength and type of lining that will be required at any given stretch of the finished tunnel.

Full mechanisation

The slow, backbreaking work of rock removal is now almost entirely mechanised. After blasting, mechanical shovels do the work of picking up the débris and loading it on to electric wagons or conveyor belts. Another method of mucking lies in the use of the shuttle car (*see* Fig. 20). Here the rock is mechanically shovelled on to a 60–70-foot unit running on rails; mechanically operated slats along the bottom of the unit move the rock back from the work-face. When the shuttle car is fully loaded it is towed out of the heading by a diesel, or an electric, locomotive.

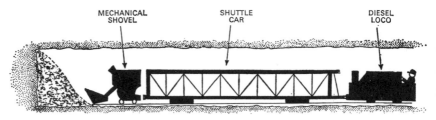

Figure 20. Mechanised rock removal.

Another method of spoil disposal is to remove it by means of a slurry-pumping plant. The spoil is mechanically loaded on to a shaker-conveyor belt that sifts it and crushes all the lumps into smaller pieces. It is then mixed with water and the resulting 'slurry' is pumped out of the tunnel through pipes. This method is especially useful for working under pressure, as the spoil does not have to pass through time-wasting air-locks. When travelling through gravelly ground, a machine can be employed to sift the gravel. It then selects the smallest pieces and, by means of compressed air, injects them through holes in the tunnel lining into the space beyond. This reduces the amount of cement required for grouting.

Using modern drills with tungsten carbide[1] bits, efficient ventilation to clear away dust and smoke together with mechanical methods of removing spoil, the drilling–blasting–mucking cycle has been reduced to as little as forty minutes, as compared with the four and a half hours that it might have taken fifty years ago (*see* p. 3). Tunnel lining may now be carried out by a machine that lifts the lining segments into position. Powered by compressed air, it has an extending arm that rotates around the axis of the heading erecting the segments into position. Hydraulic bolt-tighteners then finish the job.

Although the prototype of all mechanical tunnel-diggers was used as early as 1888 (*see* Chapter 10), until recently nearly all tunnelling was still done by the old methods. With the development of the 'drum digger', tunnellers were spared a great deal of physical toil. A development of the Greathead shield, the drum digger has an outer-shield cylinder within

[1] Tungsten carbide is an iron-grey powder produced by carbonising incandescent tungsten in a methane of hydrocarbon vapour. Highly abrasive, it is briquetted with cobalt or other binders into tools for high-speed cutting. In many cases feed and depth of cut are limited only by the ability of the machine to stand the strain.

which is a smaller cylinder, or drum, from which radiate six arms. The ends of these arms reach almost to the outer cylinder, and they are fitted with a number of renewable cutting teeth; the arms rotate with the drum, scrape against the working-face and excavate the ground. The spoil is led on to a conveyor belt that carries it back to dump it into the waiting skips. Like the Greathead shield, the drum digger advances by shoving against the previously completed tunnel lining by means of a number of powerful hydraulic jacks.

Another type of mechanical tunneller, known as the 'Mole', has a cutting head upon which are fixed fifty or more steel grinding discs, each about 1 foot in diameter. These rotate at a high speed while the whole head turns slowly. The rock is thus reduced to sand which is picked up by a series of scoops, loaded on to a conveyor and taken to the tunnel mouth. In tunnelling through hard rock that needs no lining there will be no abutment for the machine to push against; in this case the machine first drills a central hole in the work-face of about 18 inches diameter. Into this it inserts a steel rod fitted with 'spokes' that expand outward when the rock is pulled upon, but which close when the rod is pushed. Acting rather like the ribs of an umbrella, the spokes grip the rock and act as an anchor upon which the machine pulls itself forward.

As these machines are still in their infancy, there is much argument in tunnelling circles as to their relative advantages and disadvantages when compared with the conventional methods. These may be summarised as follows:

Advantages

The machine method is safer than blasting as the latter is subject to mishaps. Furthermore, blasting is liable to crack the surrounding rock, thus weakening it. The machine leaves a near-perfect circular section that is resistant to the external pressures.

Blasting breaks up to 30 per cent more rock than is necessary and this additional rock has to be loaded and carted away. Machine tunnelling reduces this wastage to 5 per cent or even less. In addition, up to 40 per cent reduction in labour costs can be achieved by the machine; less skilled labour is required. The rock is broken evenly by the machine, making it possible to lead it straight on to a conveyor belt for disposal. As the machine works continually, a better rate of advance can be made.

Disadvantages

Because of the high cost of the mechanical tunneller it can only be used for tunnels of a certain minimum length. Because of the lack of standardisation in the diameter of tunnel sections and the high cost of tunnelling machines, manufacturers do not keep a stock of them. Delivery may take up to eighteen months—a delay which is not always acceptable.

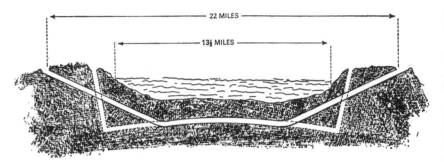

Figure 21. The Seikan Tunnel—longitudinal section, showing working shafts and pilot tunnel (not to scale).

Being in their relative infancy the machines are liable to many breakdowns. But in spite of their disadvantages, these machines have greatly increased the scope of tunnellers who now find that tunnelling under rivers and mountains is a commonplace. As a result they are turning their attention to wider horizons and it would seem that these horizons lie across the oceans. Although the first-ever under-sea tunnelling scheme, a proposed Channel tunnel, is still to be carried out, the Japanese have already made a start on a 22-mile-long tunnel under the Tsugaru straight in the Pacific Ocean, to connect Japan's main island, Honshu, to the northern island of Hokkaido. As the depth of the channel is 480 feet at its centre, the tunnel will have to slope deeply downwards from its portals— a reversal of the usual procedure for sub-aqueous tunnels and potentially a dangerous one, for if a fault in the rock is encountered the tunnel could flood under high pressure. To counter this, a pilot tunnel was first driven on an upward slope from working-shafts on each shore. This was then joined by the two main headings driven on a downward incline from the surface of each shore (*see* Fig. 21). The end sections of the pilot tunnel and its two shafts will eventually be used as a permanent drainage system. Due for completion in 1978, the Seikan tunnel, as it is known, will, undoubtedly, start an era of under-sea tunnelling which, if it develops, may eventually produce tunnels running hundreds of miles under sea-beds.

The evacuated tube railway tunnel

To overcome the problem of ventilating such immensely long tunnels, traction within them will probably be by what was known in the nineteenth century as pneumatic railway, and which is now called evacuated tube railway. A plan for a long-distance evacuated tube railway tunnel (although not an under-sea one) is at present under consideration to link the cities of Boston and Washington—a distance of 400 miles. In order to compete with the airlines it is planned to run the trains through at a cruising speed of 400 miles an hour. This speed will be attained by fitting the trains into tubes like pistons in cylinders. By evacuating the air from in front of the train by powerful pumps situated outside the tubes, the trains will be propelled forward by the pressure of the atmosphere. To provide a perfect level, it is proposed to float the tubes on water contained in the tunnel proper. It is a significant fact that such a tunnel, even though bored through hard rock, would, it is estimated, cost less in these days than an ordinary surface railway—mainly because of the high price of the land it would traverse and the present tremendous cost of bridge-building. Allowing for five intermediate stops, it is reckoned that the 400-mile journey will take ninety minutes; with six trains an hour running in both directions, up to 18,000 passengers an hour could be transported each way.

Future plans of Paul Cooper

If the Boston–Washington plan seems a little far-fetched let us consider the ideas of the American mathematician, Paul Cooper, who has calculated that tunnels could be built that would transport a passenger vehicle between any two points on the face of the earth in the space of forty-two minutes. No matter whether the journey is between Boston and Washington or London and Sydney, the journey, according to Cooper, would take the same time. He has calculated a formula for the time required for an object to fall through a straight line between any two points on earth, assuming the line to be a tunnel. The answer was a constant 42·2 minutes. A vehicle 'dropped' into a tunnel running, for instance, between London and Sydney would accelerate steadily during the first half of its trip to a speed that would produce sufficient kinetic energy to coast 'up' the other side *against* the pull of gravity to complete the journey in 42·2 minutes. Cooper's idea is an earth criss-crossed with various tunnels connecting the principal cities of the world (*see* Figs. 22, 22a).

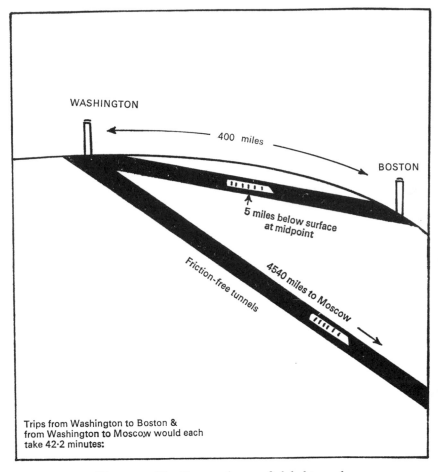

WASHINGTON

400 miles

BOSTON

5 miles below surface
at midpoint

Friction-free tunnels

4540 miles to Moscow

Trips from Washington to Boston &
from Washington to Moscow would each
take 42·2 minutes:

Figure 22. The Cooper theory of global tunnels.

With the world's cities linked by tunnels of this kind and with the departure time of all vehicles travelling through them fixed universally 'on the hour', the arrival time would be an invariable 42·2 minutes *past the hour*. This, says Cooper, will mean the abolition of all transport time-tables! As might be expected, there are one or two snags to this rather alarming new mode of travel. In the first place the tunnels would have to be kept absolutely airless and frictionless—two unobtainable conditions, without which the vehicles would be lacking sufficient kinetic energy to complete the journey. Secondly, a straight-line tunnel traversing the short distance between London and Edinburgh would, at its mid-way point, be some 5 miles below the earth's surface where the temperature might be as

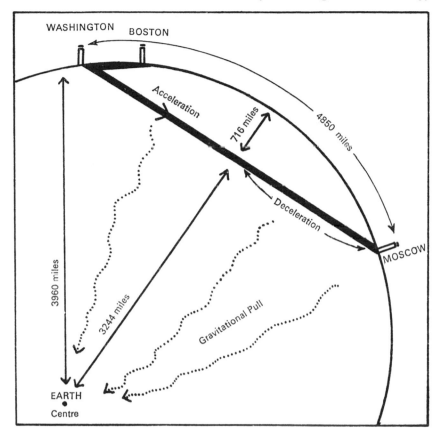

Figure 22a.

high as 260° F, requiring a passenger-carrying vehicle to be equipped with an immense cooling system. Cooper's proposed straight-line tunnel between Moscow and Washington would pass through temperatures that are unimaginable. Nevertheless, Paul Cooper is convinced that these little snags will eventually be ironed out. After all, they said James Brindley was mad!

The longest tunnel of them all

At the time of writing, a water tunnel is under construction in South Africa that will be by far the longest continuous tunnel ever constructed when it is completed in 1974. Although most of the tunnels hitherto described have been built to satisfy man's seemingly insatiable need to travel from place to place in the shortest possible time, the greatest benefits that tunnels give to the human race are those that result from one of the earliest applications of tunnelling—the underground channelling of water.

It has been said that water is South Africa's limiting factor, and such is its importance to the country's economy that the South African Government includes a Minister of Water Affairs. Over the whole country the average rainfall is only 17 inches a year; of this a mere 8 per cent reaches the river and half of this is wasted by flooding or through evaporation. Every year enough water flows down the Orange river to the sea to supply the entire needs of South Africa's most highly industrialised area, the Witwatersrand, for thirty years. In addition, 64,000 acre-feet of silt are carried into the South Atlantic and lost. With a population growing at the rate of 400,000 a year and a steadily rising standard of living for all its people, South Africa cannot afford this immense wastage of water and soil. To combat it a scheme was devised—a scheme of magnitude and grandeur that will transform a huge area of arid semi-desert into fertile land.

The Orange River Project includes the building of three huge dams; an extensive system of canals that will feed water to thousands of acres of irrigable land in the region of the Great Karoo; twenty hydro-electric stations stretching from the Indian Ocean to the South Atlantic; hundreds of miles of new roads over some of the most rugged ground in the world; six complete townships for the construction workers which will subsequently be converted into holiday centres; and a tunnel through solid rock no less than $51\frac{1}{2}$ miles long to connect the rivers Orange and Fish. The benefits of the scheme are numerous and varied. Not only will it irrigate 756,000 acres of semi-desert land, but it will supplement water supplies to existing irrigation schemes. It will make land available for 9,000 new farms; it will stimulate the production of beef, milk, wool, cotton, lucerne, wheat, raisins, beans and peas to increase the value of South African agriculture by R200,000,000 (over £100,000,000) a year. It will decentralise industry; check the drift of population from the countryside to the towns; and will stabilise the Orange river, greatly reducing, if not eliminating, the extensive damage caused by its regular floodings. A by-product will be a peak supply of 229·1 megawatts of electricity, part of which will be generated by hydro-electric generators within the tunnel itself.

Due to be completed in 1974, the tunnel has a diameter of 17 feet 6 inches and a maximum capacity of 1,120,000,000 gallons of water a day flowing down an incline of $2\frac{1}{2}$ feet to the mile. The tunnel is being advanced from sixteen work-faces, fourteen of which are provided by seven shafts ranging in depth from 250 feet to 1,245 feet (*see* Plate 17). When the tunnel is complete these shafts will be converted for permanent use for inspection and maintenance, while No. 7 shaft, the one nearest the outlet, will be used as a surge shaft, or safety valve, to control pressure fluctuations resulting from variations in the flow of water within the tunnel. The planning and investigations for the Orange–Fish tunnel began forty years ago, the geological surveys in 1948. Since then 100,000 linear feet of core have been drilled out of the ground, some of these bores having reached a depth of 1,500 feet.

The land under which the tunnel passes is known to the Hottentots as 'The Place of Great Dryness' and can only be described as a wilderness— dusty, stony and intensely hot. It was feared that such a forbidding situation and its attendant isolation would result in a rapid turnover of labour, and to avoid this it was decided to keep the standard of accommodation and living as high as possible. Three towns were built along the line of the tunnel, one at each end and one in the middle; these have electric light, purified water, water-borne sewage, tarred roads and extensive social and recreational facilities. Into these towns moved thousands of skilled engineers and workmen who came from all over the world—France, Italy, South Africa, Great Britain, Portugal, Greece, etc. Tunnelling started in March 1968 by the old method of drilling, blasting and mucking. In March 1970, the tunnellers on the inlet-portal section broke through to the heading being driven from the first vertical shaft and the concrete lining was started.

Both the headings leading from Shaft No. 2 and the shaft itself were flooded in August 1969 by an inrush of water from the face of the south heading. So great was the flooding that the advance was held up for a year in the north heading and for eighteen months in the south. Excavation was forced to a standstill in Heading 4, South, in October 1971 due to the intersection of a fissure in the rock that released methane gas which ignited. Fortunately no injuries were sustained by anyone due to either of these incidents. Seven million tons of rock will be excavated and over 2,000,000 tons of concrete poured before the Orange–Fish tunnel is finished.

Achievement of the past and future possibilities

When it is considered what has been achieved in the techniques of tunnelling since Marc Brunel invented the tunnelling-shield and burrowed under the Thames, the prospect of underground galleries hundreds of miles long is not fantastic. The average rate of advance in all tunnels driven up to 1870 has been worked out at 9 feet a week. Less than one hundred years later, in 1969, the British engineering firm of Kinncar Moodie & Co. set up a world record by tunnelling 1,215 feet in one week.

This was during the construction of an 8 feet 4 inches diameter water tunnel driven through clay with a drum digger of their own design. This record did not last long, for in the following year Kinnear Moodie's main competitor, Edmund Nuttall & Co., also driving an 8 feet 4 inches water tunnel through clay, and using one of their own machines, achieved 1,430 feet in seven days. Such a speed of advance, if it could be maintained, would make it possible to drive a 240-mile-long tunnel in ten years, if tunnelling was done from two ends—about the same period of time that it took to build the 9¼-mile-long St Gotthard tunnel in 1872–82. It is a sobering thought, however, and one which impresses on the mind just how big the earth is, that if it were possible to maintain this tunnelling speed under the bed of the Atlantic, for instance, it would take over a hundred and twenty years to advance a pilot heading (8 feet 4 inches in diameter) to stretch between Liverpool and New York (and this apart from the miles-deep working shafts that would be required). Tunnelling techniques, therefore, have a long, long way to go before such projects become feasible.

The tunnelling machine of the future will be fully automated, capable of digging, mucking, lining and, if necessary, grouting in one single coordinated operation. Laboratories are already investigating the possibilities of using other methods of penetrating ground. Hydraulic and chemical fragmentation as well as laser beams are being developed as a means of breaking rock and it is likely that the automated tunnelling machine of the future will use one or more of these methods. The age of push-button tunnelling will then have arrived.

Appendix

GENERAL NOTES AND DETAILED INSTRUCTIONS FOR COMPRESSED-AIR WORKERS

(The following is taken from a booklet issued to all compressed air-workers at the Tyne tunnel in 1965.[1])

> *The bearer of this book*
>
> . . . .
> *is a compressed-air worker.*
> *If taken ill send at once*
> *to Medical Lock.*
> EDMUND NUTTALL, SONS & CO. (LONDON) LTD.
> EAST HOWDEN,
> WALLSEND-ON-TYNE,
> *and ring EDMUND NUTTALL*
> *Wallsend 624567*

General Notes

Compressed-air chambers are used to sink shafts and drive tunnels, generally in cases where the ground is so waterlogged as to make it totally unstable.

The function of compressed air is to drive the water out of the ground and so to dry up the working face and make it manageable for the miner.

The pressure of air required depends upon the pressure of water trying to flow into the excavation. Generally speaking, the deeper the excavation the higher will be the pressure required.

[1] Reproduced by kind permission of the contractors, Edmund Nuttall Limited, who wish to make it clear that the booklet was published in March 1965 and is subject to revision in the light of any further medical research or statutory requirements made since that date.

Working in compressed air is little different from working in a free atmosphere. It is no more arduous, but for pressures over 18 lb per square inch shorter shifts are worked, usually 8 hours. Normally the limiting pressure for men working is about 50 lb per square inch, but on this job it will be less than this.

It is when workmen are passing into the chamber and passing out of it, through the man-locks, that certain precautions have to be taken.

Compression

When a person is passing into the chamber through the man-locks the gradual increase of air pressure tends to push in the ear drums. Normally this can be readily 'cleared' by frequent swallowing or by blowing the nose. Persons thus reacting normally can pass through the air-lock quite quickly with a pressure rise of up to 10 lb per square inch per minute. A person suffering from a cold in the head or with blocked ears may not, however, be able to 'clear his ears', in which case it becomes painful. **Compression must be stopped at once** and the person returned to the atmosphere. Very often after one or two attempts the person finds that the ears will 'clear', but otherwise he must not pass into the working chamber.

Decompression

In passing out from the working chamber back to atmosphere a person will suffer no inconvenience, and when the working chamber pressure has been below 18 lb per square inch decompression can be carried out at any speed with no ill effects. Above 18 lb per square inch, however, decompression must be very carefully controlled for this reason.

Ordinary air consists of a mixture of about one part of Oxygen and four parts of Nitrogen. Whilst a person has been in the working chamber the tissues of his body have been gradually absorbing an excess quantity of Nitrogen which has remained in solution under pressure. When the air pressure is reduced around him the tissues give up this Nitrogen to the atmosphere at a certain rate. If the pressure is dropped too quickly the Nitrogen cannot pass out rapidly enough to prevent bubbles of gas forming within the tissues. These will not form until some time after decompression has finished. This is known as compressed-air sickness. The symptoms vary—it may be painful and when the normal precautions have been ignored it has been known to be dangerous. The **only relief** is for the person to return to an air chamber for recompression as soon as possible. From the above it will be clear that **slow decompression** is the secret of avoiding this complaint.

For example, if a man has been working for more than 4 hours at a pressure of 25 lb per square inch, it will take about $\frac{3}{4}$ hour to decompress, and if the pressure has been 30 lb per square inch it will take about $1\frac{1}{4}$ hours.

The detailed instructions and regulations that will be issued on the works have all been framed with the object of protecting the compressed-air worker against such after effects, and are the fruits of years of research and experience. These regulations will be rigidly adhered to, and if they are faithfully complied with there should be no serious cases of 'bends'.

Detailed Instructions

1. No person will be employed on compressed-air work unless he has previously been passed fit by the Appointed Medical Officer (Dr McStay) and a certificate of fitness has been entered on his Compressed Air Health Register. This certificate will require renewing at frequent intervals to be specified by the Appointed Medical Officer.

2. Persons suffering from acute head colds or from ear, nose or throat trouble must not be employed in compressed air until specifically permitted to do so by the Appointed Medical Officer.

3. No person will be permitted to consume alcohol whilst in compressed air.

4. When the pressure exceeds 18 lb per square inch, a workman with no previous experience of compressed air shall, on his first shift, remain in the working chamber for only half the specified shift time.

5. **Man-Lock Compressions (Entering the Working Chamber)** During compressions only the Lock Keeper may delegate control of the air valves to an experienced person within the man-lock. In the event of any person suffering any pain or distress, **compression must cease at once.** Should the pain or distress persist **the lock must be returned to atmosphere and** the person released. Frequent swallowing or blowing the nose will usually relieve ear blockages and pains in the ear.

The rate of compression must not exceed 5 lb in the first minute and 10 lb per minute thereafter.

The names and numbers of all persons entering and leaving the man-lock must be logged by the Lock Keeper, together with the time of entry and leaving.

6. Man-Lock Decompressions (Leaving the Working Chamber)
All man-lock decompressions will be under the sole control of the Lock Keeper, who will be responsible for the recommended times of decompression being adhered to. It is impossible for persons within the man-lock to accelerate decompression.

7. Every person working in compressed air will be issued with an identification badge in case of sickness. This badge must be worn on the person at all times.

8. Compressed-Air Sickness
The most common form of sickness is 'bends'—acute pains in joints and muscles. These usually occur some considerable time after decompression. Generally speaking 'bends' are not dangerous, but the only effective treatment is re-compression in an air chamber as quickly as possible.

More serious forms of sickness are 'chokes' and 'staggers', and these generally occur suddenly within an hour of decompression. Immediate treatment in the medical lock is absolutely essential. Medical air-locks with skilled attendants are available on the site adjacent to the Shaft and Medical Centre.

Ambulance services have arranged to bring in men suffering from 'bends' to the Medical Centre for treatment. Any policeman or Police Station will call an ambulance.

Men who live more than 10 miles from the Site are expected to make full use of the ambulance service.

In the event of difficulty in obtaining an ambulance the cost of emergency transport will be reimbursed, but the facts must be reported to the Medical Centre on arrival.

Personal Hints

Your resistance against sickness will be substantially increased by attention to the following recommendations:

(a) Have a good meal before coming on shift and bring a substantial mid-shift snack with you.

(b) Take a warm jacket into the working chamber and wear it during decompression at the end of the shift.

(c) Use the showers and Changing Rooms and Rest Rooms provided, change into warm, dry clothes after the shift's work and stay on the site for 1 hour after decompression, when the working pressure exceeds 18/20 lb per square inch.

(d) Make yourself conversant with all signals, bells and telephones.

(e) Avoid excessive consumption of alcohol.

Bibliography

ABEL, D. *Channel Underground* (London), 1961
BLACK, A. *The Story of Tunnels* (New York), 1937
BRUNEL, I. *The Life of I. K. Brunel* (London), 1870
BRUNTON, D. W. *Modern Tunnelling* (New York), 1922
BURR, S. D. V. *Tunnelling the Hudson River* (New York), 1885
COLEMAN, T. *The Railway Navvies* (London), 1965
COWIE, H. E. C. *Tunnel work in the Punjab* (Chatham), 1905
DEAN, F. E. *Tunnels and Tunnelling* (London), 1962
DRINKER, H. S. *Tunnelling, Explosive Compounds and Rock Drills* (New York), 1878
FLAXMAN, J. *Great Feats of Engineering* (London), 1931
GIES, J. *Adventure Underground* (London), 1962
HAMMOND, R. *Tunnel Engineering* (London), 1959
HARRISON, J. L. R. *The Great Bore* (n.d.)
HARTLEY, H. A. *Famous Bridges and Tunnels of the World* (London), 1956
HAWKS, E. *Wonders of Engineering* (London), 1929
LAMPE, D. *The Tunnel* (London), 1963
L'AUCHLI, E. *Tunnelling* (New York), 1915
McMECHEN, E. G. *The Moffat Tunnel* (Denver), 1927
OVERMAN, M. *Roads, Bridges and Tunnels* (London), 1968
PEQUIGNOT, C. A. *Tunnels and Tunnelling* (New York), 1963
PRELINI, C. *Tunnelling* (New York), 1912
ROLT, L. T. C. *Isambard Kingdom Brunel* (London), 1957
SANDSTROM, E. S. *The History of Tunnelling* (London), 1963
SIMMS, F. W. *Practical Tunnelling* (London), 1877
SLATER, H. and BARNETT, C. *The Channel Tunnel* (London), 1958
SMILES, S. *Lives of the Engineers* (London), 1904
STAUFFER, D. M. *Modern Tunnel Practice* (London), 1906
SUETONIUS. *Lives of the Caesars*
Note. Useful information is also contained in the *Encyclopaedia Britannica* and the *Encyclopaedia Americana*.

Index